Collins

ROYAL
OBSERVATORY
GREENWICH

2019 GUIDE
to the
NIGHT SKY

Storm Dunlop and Wil Tirion

Published by Collins
An imprint of HarperCollins Publishers
Westerhill Road
Bishopbriggs
Glasgow G64 2QT
www.harpercollins.co.uk

In association with
Royal Museums Greenwich, the group name for the National Maritime Museum,
Royal Observatory Greenwich, Queen's House and *Cutty Sark*
www.rmg.co.uk

© HarperCollins Publishers 2018
Text and illustrations © Storm Dunlop and Wil Tirion
Photographs © see acknowledgements page 94.

A catalogue record for this book is available from the British Library

ISBN 978-0-00-825770-5

10 9 8 7 6 5 4 3 2 1

Printed in China by RR Donnelley APS

If you would like to comment on any aspect of this book, please contact us at the above address or online.
e-mail: collinsmaps@harpercollins.co.uk

 facebook.com/CollinsAstronomy

@CollinsAstro

MIX
Paper from
responsible sources
FSC™ C007454

Contents

Introduction

The aim of this Guide is to help people find their way around the night sky, by showing how the stars that are visible change from month to month and by including details of various events that occur throughout the year. The objects and events described may be observed with the naked eye, or nothing more complicated than a pair of binoculars.

The conditions for observing naturally vary over the course of the year. During the summer, twilight may persist throughout the night and make it difficult to see the faintest stars. There are three recognized stages of twilight: civil twilight, when the Sun is less than 6° below the horizon; nautical twilight, when the Sun is between 6° and 12° below the horizon; and astronomical twilight, when the Sun is between 12° and 18° below the horizon. Full darkness occurs only when the Sun is more than 18° below the horizon. During nautical twilight, only the very brightest stars are visible. During astronomical twilight, the faintest stars visible to the naked eye may be seen directly overhead, but are lost at lower altitudes. As the diagram shows, during June and most of July full darkness never occurs at the latitude of London, and at Edinburgh

nautical twilight persists throughout the whole night, so at that latitude only the very brightest stars are visible.

Another factor that affects the visibility of objects is the amount of moonlight in the sky. At Full Moon, it may be very difficult to see some of the fainter stars and objects, and even when the Moon is at a smaller phase it may seriously interfere with visibility if it is near the stars or planets in which you are interested. A full lunar calendar is given for each month and may be used to see when nights are likely to be darkest and best for observation.

The celestial sphere

All the objects in the sky (including the Sun, Moon and stars) appear to lie at some indeterminate distance on a large sphere, centred on the Earth. This *celestial sphere* has various reference points and features that are related to those of the Earth. If the Earth's rotational axis is extended, for example, it points to the North and South Celestial Poles, which are thus in line with the North and South Poles on Earth. Similarly, the *celestial equator* lies in the same plane as the Earth's equator, and divides the sky into northern and

The duration of twilight throughout the year at London and Edinburgh.

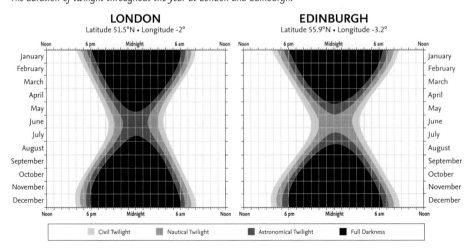

southern hemispheres. Because this Guide is written for use in Britain and Ireland, the area of the sky that it describes includes the whole of the northern celestial hemisphere and those portions of the southern that become visible at different times of the year. Stars in the far south, however, remain invisible throughout the year, and are not included.

It is useful to know some of the special terms for various parts of the sky. As seen by an observer, half of the celestial sphere is invisible, below the horizon. The point directly overhead is known as the **zenith**, and the (invisible) one below one's feet as the **nadir**. The line running from the north point on the horizon, up through the zenith and then down to the south point is the **meridian**. This is an important invisible line in the sky, because objects are highest in the sky, and thus easiest to see, when they cross the meridian in the south. Objects are said to **transit**, when they cross this line in the sky.

In this book, reference is frequently made in the text and in the diagrams to the standard compass points around the horizon. The position of any object in the sky may be described by its **altitude** (measured in degrees above the horizon), and its **azimuth** (measured

in degrees from north 0°, through east 90°, south 180° and west 270°). Experienced amateurs and professional astronomers also use another system of specifying locations on the celestial sphere, but that need not concern us here, where the simpler method will suffice.

The celestial sphere appears to rotate about an invisible axis, running between the North and South Celestial Poles. The location (i.e., the altitude) of the Celestial Poles depends entirely on the observer's position on Earth or, more specifically, their latitude. The charts in this book are produced for the latitude of 50°N, so the North Celestial Pole (NCP) is 50° above the northern horizon. The fact that the NCP is fixed relative to the horizon means that all the stars within 50° of the pole are always above the horizon and may, therefore, always be seen at night, regardless of the time of year. This northern circumpolar region is an ideal place to begin learning the sky, and ways to identify the circumpolar stars and constellations will be described shortly.

The ecliptic and the zodiac

Another important line on the celestial sphere is the Sun's apparent path against the background stars – in reality the result

Measuring altitude and azimuth on the celestial sphere.

The altitude of the North Celestial Pole equals the observer's latitude.

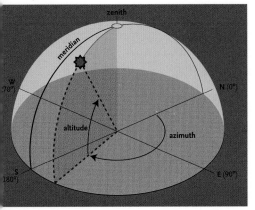

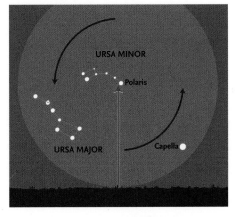

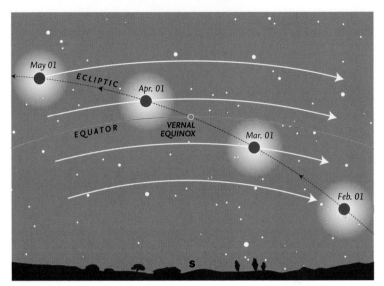

The Sun crossing the celestial equator in spring.

of the Earth's orbit around the Sun. This is known as the **ecliptic**. The point where the Sun, apparently moving along the ecliptic, crosses the celestial equator from south to north is known as the vernal (or spring) equinox, which occurs on March 20. At this time (and at the autumnal equinox, on September 22 or 23, when the Sun crosses the celestial equator from north to south) day and night are almost exactly equal in length. (There is a slight difference, but that need not concern us here.) The vernal equinox is currently located in the constellation of Pisces, and is important in astronomy because it defines the zero point for a system of celestial coordinates, which is, however, not used in this Guide.

The Moon and planets are to be found in a band of sky that extends 8° on either side of the ecliptic. This is because the orbits of the Moon and planets are inclined at various angles to the ecliptic (i.e., to the plane of the Earth's orbit). This band of sky is known as the zodiac and, when originally devised, consisted of twelve **constellations**, all of which were considered to be exactly 30° wide. When the constellation boundaries were formally established by the International Astronomical

Union in 1930, the exact extent of most constellations was altered and, nowadays, the ecliptic passes through thirteen constellations. Because of the boundary changes, the Moon and planets may actually pass through several other constellations that are adjacent to the original twelve.

The constellations

Since ancient times, the celestial sphere has been divided into various constellations, most dating back to antiquity and usually associated with certain myths or legendary people and animals. Nowadays, the boundaries of the constellations have been fixed by international agreement and their names (in Latin) are largely derived from Greek or Roman originals. Some of the names of the most prominent stars are of Greek or Roman origin, but many are derived from Arabic names. Many bright stars have no individual names and, for many years, stars were identified by terms such as 'the star in Hercules' right foot'. A more sensible scheme was introduced by the German astronomer Johannes Bayer in the early seventeenth century. Following his scheme – which is still used today – most of the brightest stars are identified by a Greek letter followed by the genitive form of the constellation's Latin name. An example is the Pole Star, also known as Polaris and α Ursae Minoris (abbreviated α UMi). The Greek alphabet is shown on page 93 with a list of all the constellations that may be seen from latitude 50°N, together with abbreviations, their genitive forms and English names. Other naming schemes exist for fainter stars, but are not used in this book.

Asterisms

Apart from the constellations (88 of which cover the whole sky), certain groups of stars, which may form a part of a larger constellation or cross several constellations, are readily recognizable and have been given individual names. These groups are known as *asterisms*, and the most famous (and well-known) is the 'Plough', the common name for the seven brightest stars in the constellation of Ursa Major, the Great Bear. The names and details of some asterisms mentioned in this book are given in the list on page 94.

Magnitudes

The brightness of a star, planet or other body is frequently given in magnitudes (mag.). This is a mathematically defined scale where larger numbers indicate a fainter object. The scale extends beyond the zero point to negative numbers for very bright objects. (Sirius, the brightest star in the sky is mag. -1.4.) Most observers are able to see stars to about mag. 6, under very clear skies.

The Moon

Although the daily rotation of the Earth carries the sky from east to west, the Moon gradually moves eastwards by approximately its diameter (about half a degree) in an hour. Normally, in its orbit around the Earth, the Moon passes above or below the direct line between Earth and Sun (at New Moon) or outside the area obscured by the Earth's shadow (at Full Moon). Occasionally, however, the three bodies are more-or-less perfectly aligned to give an

eclipse: a solar eclipse at New Moon or a lunar eclipse at Full Moon. Depending on the exact circumstances, a solar eclipse may be merely partial (when the Moon does not cover the whole of the Sun's disk); annular (when the Moon is too far from Earth in its orbit to appear large enough to hide the whole of the Sun); or total. Total and annular eclipses are visible from very restricted areas of the Earth, but partial eclipses are normally visible over a wider area.

Somewhat similarly, at a lunar eclipse, the Moon may pass through the outer zone of the Earth's shadow, the *penumbra* (in a penumbral eclipse, which is not generally perceptible to the naked eye), so that just part of the Moon is within the darkest part of the Earth's shadow, the *umbra* (in a partial eclipse); or completely within the umbra (in a total eclipse). Unlike solar eclipses, lunar eclipses are visible from large areas of the Earth.

Occasionally, as it moves across the sky, the Moon passes between the Earth and individual planets or distant stars, giving rise to an *occultation*. As with solar eclipses, such occultations are visible from restricted areas of the world.

The planets

Because the planets are always moving against the background stars, they are treated in some detail in the monthly pages and information is given when they are close to other planets, the Moon or any of five bright stars that lie near the ecliptic. Such events are known as *appulses* or, more frequently, as *conjunctions*. (There are technical differences in the way these terms are defined – and should be used – in astronomy, but these need not concern us here.) The positions of the planets are shown for every month on a special chart of the ecliptic.

The term conjunction is also used when a planet is either directly behind or in front of the Sun, as seen from Earth. (Under normal circumstances it will then be invisible.) The conditions of most favourable visibility depend on whether the planet is one of the two known

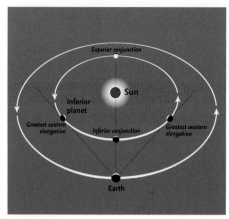

Inferior planet.

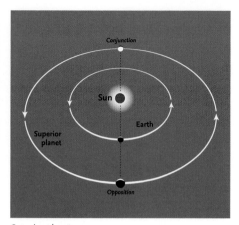

Superior planet.

as **inferior planets** (Mercury and Venus) or one of the three **superior planets** (Mars, Jupiter and Saturn) that are covered in detail. (Some details of the fainter superior planets, Uranus and Neptune, are included in this Guide, and special charts for both are given on pages 75 and 71.)

The inferior planets are most readily seen at eastern or western **elongation**, when their angular distance from the Sun is greatest. For superior planets, they are best seen at **opposition**, when they are directly opposite the Sun in the sky, and cross the meridian at local midnight.

It is often useful to be able to estimate angles on the sky, and approximate values may be obtained by holding one hand at arm's length. The various angles are shown in the diagram, together with the separations of the various stars in the Plough.

Meteors

At some time or other, nearly everyone has seen a **meteor** – a 'shooting star' – as it flashed across the sky. The particles that cause meteors – known technically as 'meteoroids' – range in size from that of a grain of sand (or even smaller) to the size of a pea. On any night of the year there are occasional meteors, known as **sporadics**, that may travel in any direction. These occur at a rate that is normally between three and eight in an hour. Far more

important, however, are **meteor showers**, which occur at fixed periods of the year, when the Earth encounters a trail of particles left behind by a comet or, very occasionally, by a minor planet (asteroid). Meteors always appear to diverge from a single point on the sky, known as the **radiant**, and the radiants of major showers are shown on the charts. Meteors that come from a circular area 8° in diameter around the radiant are classed as belonging to the particular shower. All others that do not come from that area are sporadics (or, occasionally from another shower that is active at the same time). A list of the major meteor showers is given on page 17.

Although the positions of the various shower radiants are shown on the charts, looking directly at the radiant is not the most effective way of seeing meteors. They are most likely to be noticed if one is looking about 40–45° away from the radiant position. (This is approximately two hand-spans as shown in the diagram for measuring angles.)

Other objects

Certain other objects may be seen with the naked eye under good conditions. Some were given names in antiquity – Praesepe is one example – but many are known by what are called 'Messier numbers', the numbers in a catalogue of nebulous objects compiled by Charles Messier in the late eighteenth century.

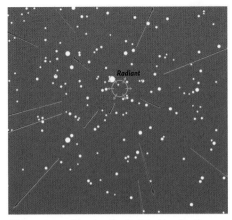

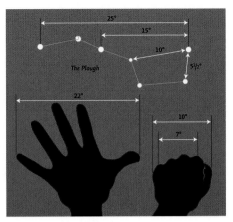

Meteor shower (showing the April Lyrid radiant).

Measuring angles in the sky.

Some, such as the Andromeda Galaxy, M31, and the Orion Nebula, M42, may be seen by the naked eye, but all those given in the list will benefit from the use of binoculars. Apart from galaxies, such as M31, which contain thousands of millions of stars, there are also two types of cluster: open clusters, such as M45, the Pleiades, which may consist of a few dozen to some hundreds of stars; and globular clusters, such as M13 in Hercules, which are spherical concentrations of many thousands of stars. One or two gaseous nebulae, consisting of gas illuminated by stars within them, are also visible. The Orion Nebula, M42, is one, and is illuminated by the group of four stars, known as the Trapezium, which may be seen within it by using a good pair of binoculars.

Some interesting objects.

Messier / NGC	Name	Type	Constellation	Maps (months)
—	Hyades	open cluster	Taurus	Sep. – Apr.
—	Double Cluster	open cluster	Perseus	All year
—	Melotte 111 (Coma Cluster)	open cluster	Coma Berenices	Jan. – Aug.
M3	—	globular cluster	Canes Venatici	Jan. – Sep.
M4	—	globular cluster	Scorpius	May – Aug.
M8	Lagoon Nebula	gaseous nebula	Sagittarius	Jun. – Sep.
M11	Wild Duck Cluster	open cluster	Scutum	May – Oct.
M13	Hercules Cluster	globular cluster	Hercules	Feb. – Nov.
M15	—	globular cluster	Pegasus	Jun. – Dec.
M22	—	globular cluster	Sagittarius	Jun. – Sep.
M27	Dumbbell Nebula	planetary nebula	Vulpecula	May – Dec.
M31	Andromeda Galaxy	galaxy	Andromeda	All year
M35	—	open cluster	Gemini	Oct. – May
M42	Orion Nebula	gaseous nebula	Orion	Nov. – Mar.
M44	Praesepe	open cluster	Cancer	Nov. – Jun.
M45	Pleiades	open cluster	Taurus	Aug. – Apr.
M57	Ring Nebula	planetary nebula	Lyra	Apr. – Dec.
M67	—	open cluster	Cancer	Dec. – May
NGC 752	—	open cluster	Andromeda	Jul. – Mar.
NGC 3242	Ghost of Jupiter	planetary nebula	Hydra	Feb. – May

The Northern Circumpolar Constellations

The northern circumpolar stars are the key to starting to identify the constellations. For anyone in the northern hemisphere they are visible at any time of the year, and nearly everyone is familiar with the seven stars of the Plough – known as the Big Dipper in North America – an asterism that forms part of the large constellation of **Ursa Major** (the Great Bear).

Ursa Major

Because of the movement of the stars caused by the passage of the seasons, Ursa Major lies in different parts of the evening sky at different periods of the year. The diagram below shows its position for the four main seasons. The seven stars of the Plough remain visible throughout the year anywhere north of latitude 40°N. Even at the latitude (50°N) for which the charts in this book are drawn, many of the stars in the southern portion of the constellation of Ursa Major are hidden below the horizon for part of the year or (particularly in late summer) cannot be seen late in the night.

Polaris and Ursa Minor

The two stars **Dubhe** and **Merak** (α and β Ursae

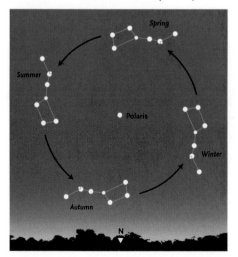

Majoris, respectively), farthest from the 'tail' are known as the 'Pointers'. A line from Merak to Dubhe, extended about five times their separation, leads to the Pole Star, **Polaris**, or α Ursae Minoris. All the stars in the northern sky appear to rotate around it. There are five main stars in the constellation of **Ursa Minor**, and the two farthest from the Pole, **Kochab** and **Pherkad** (β and γ Ursae Minoris, respectively), are known as 'The Guards'.

Cassiopeia

On the opposite of the North Pole from Ursa Major lies **Cassiopeia**. It is highly distinctive, appearing as five stars forming a letter 'W' or 'M' depending on its orientation. Provided the sky is reasonably clear of clouds, you will nearly always be able to see either Ursa Major or Cassiopeia, and thus be able to orientate yourself on the sky.

To find Cassiopeia, start with **Alioth** (ε Ursae Majoris), the first star in the tail of the Great Bear. A line from this star extended through Polaris points directly towards γ Cassiopeiae, the central star of the five.

Cepheus

Although the constellation of **Cepheus** is fully circumpolar, it is not nearly as well-known as Ursa Major, Ursa Minor or Cassiopeia, partly because its stars are fainter. Its shape is rather like the gable end of a house. The line from the Pointers through Polaris, if extended, leads to **Errai** (γ Cephei) at the 'top' of the 'gable'. The brightest star, **Alderamin** (α Cephei) lies in the Milky Way region, at the 'bottom right-hand corner' of the figure.

Draco

The constellation of **Draco** consists of a quadrilateral of stars, known as the 'Head of Draco' (and also the 'Lozenge'), and a long chain of stars forming the neck and body of the dragon. To find the Head of Draco, locate the two stars **Phecda** and **Megrez** (γ and δ Ursae Majoris) in the Plough, opposite the Pointers.

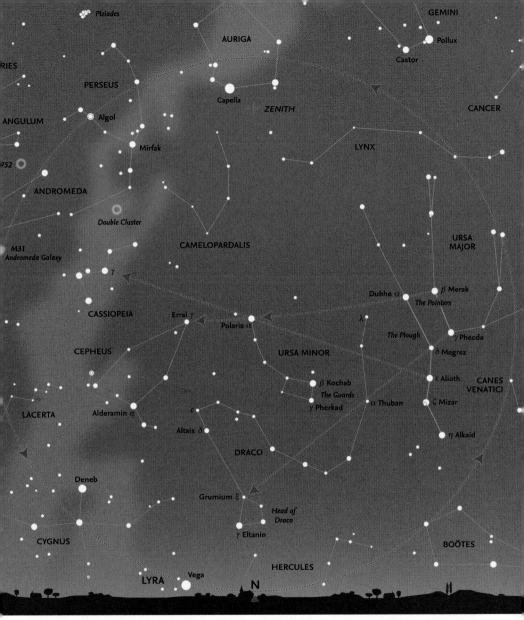

The stars and constellations inside the circle are always above the horizon, seen from our latitude.

Extend a line from Phecda through Megrez by about eight times their separation, right across the sky below the Guards in Ursa Minor, ending at **Grumium** (ξ Draconis) at one corner of the quadrilateral. The brightest star, **Eltanin** (γ Draconis) lies farther to the south. From the head of Draco, the constellation first runs northeast to **Altais** (δ Draconis) and ε Draconis, then doubles back southwards before winding its way through **Thuban** (α Draconis) before ending at λ Draconis between the Pointers and Polaris.

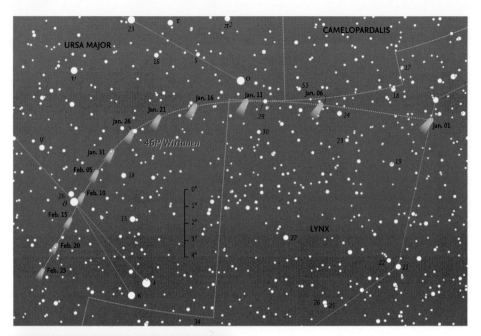

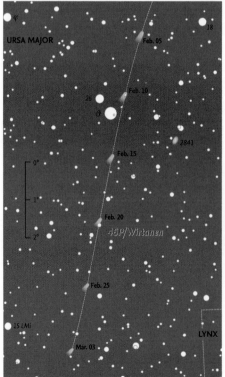

The path of comet 46P/Wirtanen from January 1 to March 3, 2019.

The chart above shows stars down to magnitude 8.5 and the chart on the left shows stars to magnitude 10.0.

Comet C/2014 Q2 Lovejoy, which reached naked-eye visibility, photographed on 20 December 2014, when in Columba, by Damian Peach.

Comets and the Moon

Comet C/2006 P1 McNaught, imaged on 20 January 2007, from Lawlers Gold Mine, Western Australia (Photographer: Sjbmgrtl).

Comets

Although comets may occasionally become very striking objects in the sky, their occurrence and particularly the existence or length of any tail and their overall magnitude are notoriously difficult to predict. Naturally, it is only possible to predict the return of periodic comets (whose names have the prefix 'P'). Many comets appear unexpectedly (these have names with the prefix 'C'). Bright, readily visible comets such as C/1995 Y1 Hyakutake & C/1995 O1 Hale-Bopp or C/2006 P1 McNaught (sometimes known as the Great Comet of 2007) are rare. Comet Hale-Bopp, in particular, was visible for a record 18 months and was a prominent object in northern skies. Comet McNaught was notable for its multiple tail structure. Most periodic comets are faint and only a very small number ever become bright enough to be readily visible with the naked eye or with binoculars. One comet, 46P/Wirtanen, was expected to become visible in binoculars in October 2018. It is well placed in the northern sky in the first few months of 2019. The accompanying charts show its path at its brightest during early 2019 until March, when it is expected to fade below magnitude 10.

The Moon

The monthly pages include diagrams showing the phase of the Moon for every day of the month, and also indicate the day in the *lunation* (or *age* of the Moon), which begins at New Moon. Although the main features of the surface – the light highlands and the dark maria (seas) – may be seen with the naked eye, far more features may be detected with the use of binoculars or any telescope. The many craters are best seen when they are close to the *terminator* (the boundary between the illuminated and the non-illuminated areas of the surface), when the Sun rises or sets over any particular region of the Moon and the crater walls or central peaks cast strong shadows. Most features become difficult to see at Full Moon, although this is the best time to see the bright ray systems surrounding certain craters. Accompanying the Moon map on the following pages is a list of prominent features, including the days in the lunation when features are normally close to the terminator and thus easiest to see. A few bright features such as Linné and Proclus, visible when well illuminated, are also listed. One feature, Rupes Recta (the Straight Wall) is readily visible only when it casts a shadow with light from the east, appearing as a light line when illuminated from the opposite direction.

The dates of visibility vary slightly through the effects of *libration*. Because the Moon's orbit is inclined to the Earth's equator and also because it moves in an ellipse, the Moon appears to rock slightly from side to side (and nod up and down). Features near the *limb* (the edge of the Moon) may vary considerably in their location and visibility. (This is easily noticeable with Mare Crisium and the craters Tycho and Plato.) Another effect is that at crescent phases before and after New Moon, the normally non-illuminated portion of the Moon receives a certain amount of light, reflected from the Earth. This *Earthshine* may enable certain bright features (such as Aristarchus, Kepler and Copernicus) to be detected even though they are not illuminated by sunlight.

Map of the Moon

NORTH

Goldschmidt
Philolaus
Pythagoras · Carpenter
Barrow
· J. Herschel
W. Bond
60°

MARE FRIGORIS

Harpalus · Plato
SINUS RORIS
Bianchini
Vallis Alpes
Sharp · Montes Jura
Montes Recti
Mons Pico
Montes Alpes
SINUS IRIDIUM
Mairan · Mons Rümker
Helicon · Le Verrier
Cassini

MARE IMBRIUM
Aristillus
Autolycus
30° Delisle
Archimedes

PALUS PUTREDINIS
Vallis Schröteri · Prinz
Timocharis
Lambert
Montes Apenninus
· Struve · Aristarchus Herodotus
· Euler
Cocon
· Seleucus

· Krafft
Montes Carpatus
MARE VAPORUM
· Cardanus
Mayer · Gay-Lussac
· Eratosthenes
· Marius
SINUS AESTUUM
Copernicus · Stadius
Kepler
· Reiner
Bode · Triesnecker
· Encke
· Cavalerius
· Reinhold
SINUS MEDII
W · Hevelius
OCEANUS PROCELLARUM
Landsberg · Gambart
60° 30° 0°
Mösting ·
Flammarion ·
Grimaldi
Montes Riphaeus
Fra Mauro ·
Herschel ·
· Parry
· Hansteen · Letronne
MARE COGNITUM
Bonpland
Ptolemaeus ·
Sirsalis A · Sirsalis
Guericke ·
Davy ·
Albategnius
· Billy
Alphonsus
· Gassendi
MARE NUBIUM
Alpetragius ·
Agatharchides
Arzachel ·
· Mersenius
· Bullialdus
Birt · Thebit
La Caille
· Byrgius
MARE HUMORUM
Campanus
Purbach Blanchinus
· Vieta Vitello
Mercator Pitatus
Regiomontanus Werner
PALUS EPIDEMIARUM
Gauricus
Walther · Aliacensis
Cichus Wurzelbauer
Nonius
Capuanus
Heisius ·
Orontius
Stöfler
Mee ·
Wilhelm
· Nasireddin
· Schickard
Tycho
Saussure
Licetus
Wargentin
Maginus
Phocylides Schiller
Longomontanus
Clavius ·
60°
Scheiner
Curtius
Blancanus
Moretus ·

SOUTH

14

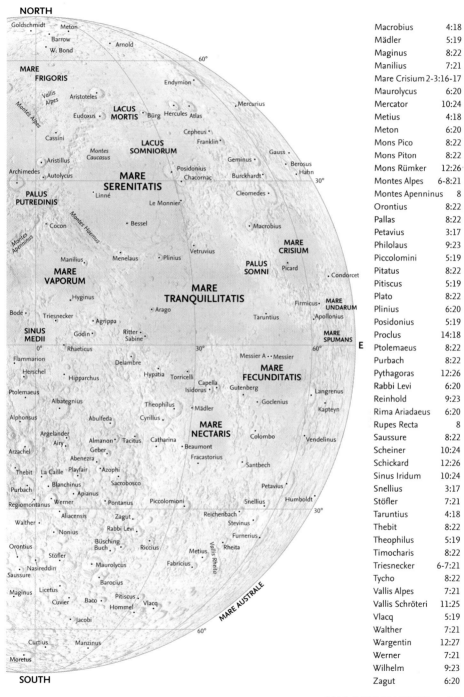

Introduction to the Month-by-Month Guide

The monthly charts

The pages devoted to each month contain a pair of charts showing the appearance of the night sky, looking north and looking south. The charts (as with all the charts in this book) are drawn for the latitude of 50°N, so observers farther north will see slightly more of the sky on the northern horizon, and slightly less on the southern. These areas are, of course, those most likely to be affected by poor observing conditions caused by haze, mist or smoke. In addition, stars close to the horizon are always dimmed by atmospheric absorption, so sometimes the faintest stars marked on the charts may not be visible.

The three times shown for each chart require a little explanation. The charts are drawn to show the appearance at 23:00 GMT for the 1st of each month. The same appearance will apply an hour earlier (22:00 GMT) on the 15th, and yet another hour earlier (21:00 GMT) at the end of the month (shown as the 1st of the following month). GMT is identical to the Universal Time (UT) used by astronomers around the world. In Europe, Summer Time is introduced in March, so the March charts apply to 23:00 GMT on March 1, 22:00 GMT on March 15, but 22:00 BST (British Summer Time) on April 1. The change back from Summer Time (in Europe) occurs in October, so the charts for that month apply to 00:00 BST for October 1, 23:00 BST for October 15, and 21:00 GMT for November 1.

The charts may be used for earlier or later times during the night. To observe two hours earlier, use the charts for the preceding month; for two hours later, the charts for the next month.

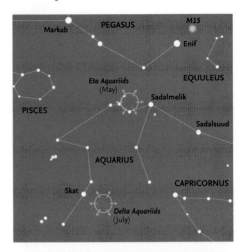

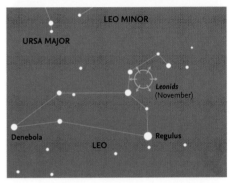

Meteors

Details of specific meteor showers are given in the months when they come to maximum, regardless of whether they begin or end in other months. Note that not all the respective radiants are marked on the charts for that particular month, because the radiants may be below the horizon, or lie in constellations that are not readily visible during the month of maximum. For this reason, special charts for the Eta and Delta Aquariids (May and July, respectively) and the Leonids (November) are given here. As explained earlier, however, meteors from such showers may still be seen, because the most effective region for seeing meteors is some 40–45° away from the radiant, and that area of sky may well be above the horizon. A table of the best meteor showers visible during the year is also given here.

Meteors that are brighter than magnitude -4 (approximately the maximum magnitude reached by Venus) are known as *fireballs* or

Shower	Dates of activity 2019	Date of maximum 2019	Possible hourly rate
Quadrantids	January 1–10	January 3–4	120
April Lyrids	April 16–25	April 21–22	18
Eta Aquariids	April 19 to May 26	May 6–7	55
Alpha Capricornids	July 11 to August 10	July 26 to August 1	5
Perseids	July 13 to August 26	August 11–12	100
Delta Aquariids	July 21 to August 23	July 29–30	< 20
Alpha Aurigids	August to October	August 28 & September 15	10
Southern Taurids	September 23 to November 19	October 28–29	< 5
Orionids	September 23 to November 27	October 21–22	25
Northern Taurids	October 19 to December 10	November 10–11	< 5
Leonids	November 5–30	November 17–18	< 15
Geminids	December 4–16	December 13–14	100+
Ursids	December 17–23	December 21–22	< 10

bolides. Examples are shown on pages 23 and 63. Fireballs sometimes cause sonic booms that may be heard some time after the meteor is seen.

The photographs

As an aid to identification – especially as some people find it difficult to relate charts to the actual stars they see in the sky – one or more photographs of constellations visible in certain specific months are included. It should be noted, however, that because of the limitations of the photographic and printing processes, and the differences between the sensitivity of different individuals to faint starlight (especially in their ability to detect different colours), and the degree to which they have become adapted to the dark, the apparent brightness of stars in the photographs will not necessarily precisely match that seen by any one observer.

The Moon calendar

The Moon calendar is largely self-explanatory. It shows the phase of the Moon for every day of the month, with the exact times (in Universal Time) of New Moon, First Quarter, Full Moon and Last Quarter. Because the times are calculated from the Moon's actual orbital parameters, some of the times shown will, naturally, fall during daylight, but any difference is too small to affect the appearance of the Moon on that date. Also shown is the **age** of the Moon (the day in the **lunation**),

beginning at New Moon, which may be used to determine the best time for observation of specific lunar features.

The Moon

The section on the Moon includes details of any lunar or solar eclipses that may occur during the month (visible from anywhere on Earth). Similar information is given about any important occultations. Mainly, however, this section summarizes when the Moon passes close to planets or the five prominent stars close to the ecliptic. The dates when the Moon is closest to the Earth (at **perigee**) and farthest from it (at **apogee**) are shown in the monthly calendars, and only mentioned here when they are particularly significant, such as the nearest and farthest during the year.

The planets and minor planets

Brief details are given of the location, movement and brightness of the planets from Mercury to Saturn throughout the month. None of the planets can, of course, be seen when they are close to the Sun, so such periods are generally noted. All of the planets may sometimes lie on the opposite side of the Sun to the Earth (at superior conjunction), but in the case of the inferior planets, Mercury and Venus, they may also pass between the Earth and the Sun (at inferior conjunction) and are invisible for a period of time, the length of which varies from conjunction to conjunction.

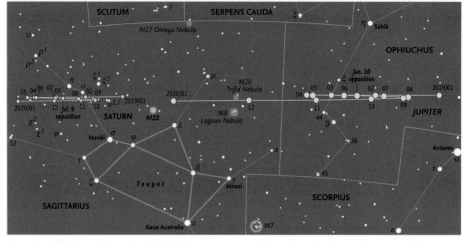

The paths of Jupiter and Saturn in 2019. Jupiter comes to opposition on June 10; Saturn almost a month later, on July 9. Background stars are shown down to magnitude 6.5.

Those two planets are normally easiest to see around either eastern or western elongation, in the evening or morning sky, respectively. Not every elongation is favourable, so although every elongation is listed, only those where observing conditions are favourable are shown in the individual diagrams of events.

The dates at which the superior planets reverse their motion (from direct motion to **retrograde**, and retrograde to direct) and of opposition (when a planet generally reaches its maximum brightness) are given. Some planets, especially distant Saturn, may spend most or all of the year in a single constellation. Jupiter and Saturn are normally easiest to see around opposition, which occurs every year. Mars, by contrast, moves relatively rapidly against the background stars and in some years never comes to opposition.

Uranus is not always included in the monthly details because it is generally at the limit of naked-eye visibility (magnitude 5.7–5.9), although bright enough to be visible in binoculars, or even with the naked eye under exceptionally dark skies. Its path in 2019 is shown on the special chart in October. It comes to opposition on October 28 in the constellation of Aries. New Moon occurs that day, so the planet, at magnitude 5.7 should

be detectable reasonably easily. It is at the same magnitude for an extended period of the year (from August to December) and should be visible when free from interference by moonlight.

Similar considerations apply to Neptune, although this is always fainter (magnitude 7.8–8.0 in 2019), but still visible in most binoculars. It reaches opposition at mag. 7.8 on September 10 in Aquarius. Full Moon occurs four days later, so Neptune will be difficult to detect, for about a week after Full Moon. Again, Neptune's path in 2019 and its position at opposition are shown in a chart in September.

Charts for the three brightest minor planets that come to opposition in 2019 are shown in the relevant month: Pallas (mag. 7.9) on April 10, Ceres (mag.7.0) on May 28, and Vesta (mag. 6.5) on November 12.

The ecliptic charts

Although the ecliptic charts are primarily designed to show the positions and motions of the major planets, they also show the motion of the Sun during the month. The light-tinted area shows the area of the sky that is invisible during daylight, but the darker area gives an indication of which constellations are likely to be visible at some time of the night.

The closer a planet is to the border between dark and light, the more difficult it will be to see in the twilight.

The monthly calendar

For each month, a calendar shows details of significant events, including when planets are close to one another in the sky, close to the Moon, or close to any one of five bright stars that are spaced along the ecliptic. The times shown are given in Universal Time (UT), always used by astronomers throughout the year, and which is identical to Greenwich Mean Time (GMT). So during the summer months, they do not show Summer Time, which will always be one hour later than the time shown.

The diagrams of interesting events

Each month, a number of diagrams show the appearance of the sky when certain events take place. However, the exact positions of celestial objects and their separations greatly depend on the observer's position on Earth. When the Moon is one of the objects involved, because it is relatively close to Earth, there may be very significant changes from one location to another. Close approaches between planets or between a planet and a star are less affected by changes of location, which may thus be ignored.

The diagrams showing the appearance of the sky are drawn for the latitude of London, so will be approximately correct for most of Britain and Europe. However, for an observer farther north (say Edinburgh), a planet or star listed as being north of the Moon will appear even farther north, whereas one south of the Moon will appear closer to it – or may even be hidden (occulted) by it. For an observer farther south than London, there will be corresponding changes in the opposite direction: for a star or planet south of the Moon the separation will increase, and for one north of the Moon the separation will decrease. This is particularly important when occultations occur, which may be visible from one location, but not another. However, there are no major occultations visible from Britain in 2019.

Ideally, details should be calculated for each individual observer, but this is obviously impractical. In fact, positions and separations are actually calculated for a theoretical observer located at the centre of the Earth.

So the details given regarding the positions of the various bodies should be used as a guide to their location. A similar situation arises with the times that are shown. These are calculated according to certain technical criteria, which need not concern us here. However, they do not necessarily indicate the exact time when two bodies are closest together. Similarly, dates and times are given, even if they fall in daylight, when the objects are likely to be completely invisible. However, such times do give an indication that the objects concerned will be in the same general area of the sky during both the preceding, and the following nights.

Key to the symbols used on the monthy star maps.

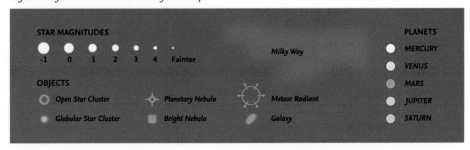

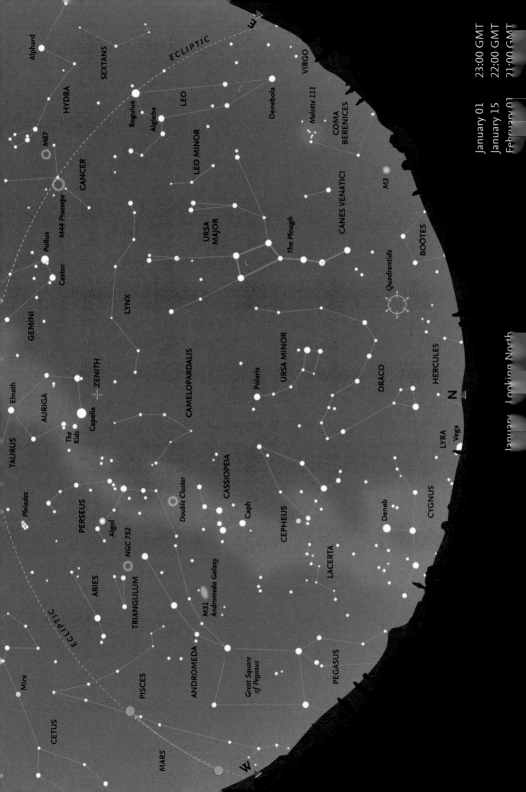

ECLIPTIC

Alphard

SEXTANS

HYDRA

M67

CANCER

M44 Praesepe

Pollux

Castor

GEMINI

Elnath

AURIGA

The Kids

Capella

TAURUS

+ ZENITH

Pleiades

PERSEUS

Algol

NGC 752

ARIES

TRIANGULUM

Mira

CETUS

ECLIPTIC

PISCES

MARS

ANDROMEDA

M31
Andromeda Galaxy

Great Square
of Pegasus

PEGASUS

LACERTA

Double Cluster

CASSIOPEIA

Caph

CEPHEUS

Deneb

CYGNUS

LYRA

Vega

CAMELOPARDALIS

LYNX

Regulus

Algieba

LEO

LEO MINOR

URSA MAJOR

The Plough

Polaris

URSA MINOR

DRACO

HERCULES

N

Denebola

VIRGO

Melotte 111

COMA
BERENICES

M3

CANES VENATICI

BOÖTES

Quadrantids

W

January – Looking North

Most of the important circumpolar constellations are easy to see in the northern sky at this time of year. **Ursa Major** stands more-or-less vertically above the horizon in the northeast, with the zodiacal constellation of **Leo** rising in the east. To the north, the stars of **Ursa Minor** lie below **Polaris** (the Pole Star). The head of **Draco** is low on the northern horizon, but may be difficult to see unless observing conditions are good. Both **Cepheus** and **Cassiopeia** are readily visible in the northwest, and even the faint constellation of **Camelopardalis** is high enough in the sky for it to be easily visible.

Near the zenith is the constellation of **Auriga** (the Charioteer), with brilliant **Capella** (α Aurigae), directly overhead. Slightly to the west of Capella lies a small triangle of fainter stars, known as 'The Kids.' (Ancient mythological representations of Auriga show him carrying two young goats.) Together with the northernmost bright star in Taurus, **Elnath** (β Tauri), the body of Auriga forms a large pentagon on the sky, with The Kids lying on the western side. Farther down towards the west are the constellations of **Perseus** and **Andromeda**, and the Great Square of **Pegasus** is approaching the horizon.

Meteors

One of the strongest and most consistent meteor showers of the year occurs in January: the **Quadrantids,** which are visible January 1–10, with maximum on January 3–4. They are brilliant, bluish and yellowish-white meteors and fireballs (page 23), which at maximum may even reach a rate of 120 meteors per hour. At maximum, the Moon is a waning crescent, just before New Moon, so there will be little interference from moonlight. The parent object is minor planet 2003 EH_1.

The shower is named after the former constellation **Quadrans Muralis** (the Mural Quadrant), an early form of astronomical

The constellation of Orion dominates the sky during this period of the year, and is a useful starting point for recognizing other constellations in the southern sky. Here, orange Betelgeuse, blue-white Rigel and the pinkish Orion Nebula are prominent. Orion can be found in the southern part of the sky (see next page).

instrument. The Quadrantid meteor radiant is now within the northernmost part of **Boötes,** roughly halfway between θ Boötis and τ Herculis.

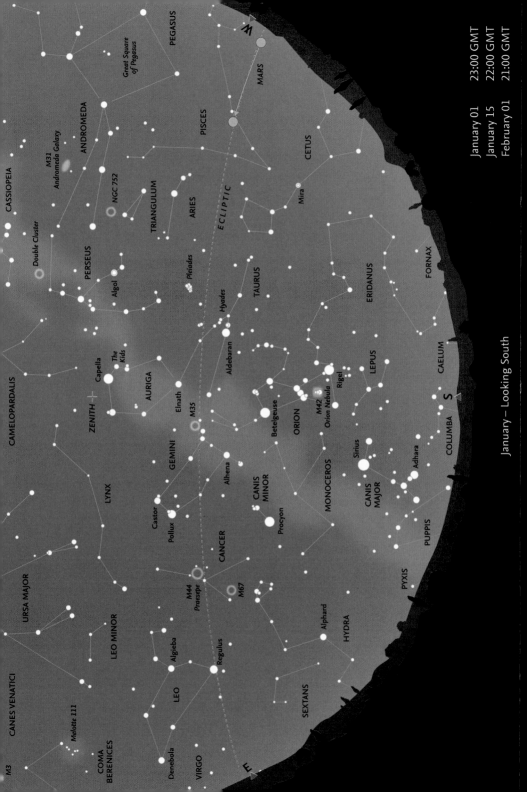

January – Looking South

January 01   23:00 GMT
January 15   22:00 GMT
February 01   21:00 GMT

January – Looking South

At this time of year the southern sky is dominated by **Orion.** This is the most prominent constellation during the winter months, when it is visible at some time during the night. (A photograph of Orion appears on page 21.) It has a highly distinctive shape, with a line of three stars that form the 'Belt'. To most observers, the bright star at the northeastern corner of the constellation, **Betelgeuse** (α Orionis), shows a reddish tinge, in contrast to the brilliant bluish-white colour of the bright star at the southwestern corner, **Rigel** (β Orionis). The three stars of the belt lie directly south of the celestial equator. A vertical line of three 'stars' forms the 'Sword' that hangs to the south of the Belt. With good viewing conditions, the central 'star' appears as a hazy spot, even to the naked eye. This is actually the **Orion Nebula.** Binoculars will reveal the four stars of the '**Trapezium**', which illuminate the nebula.

The line of Orion's Belt points up to the northwest towards **Taurus** (the Bull) and below orange-tinted **Aldebaran** (α Tauri). Close to Aldebaran, there is a conspicuous 'V' of stars, pointing down to the southwest, called the **Hyades** cluster. (Despite appearances, Aldebaran is not part of the cluster.) Farther along, the same line from Orion

A late Quadrantid fireball, photographed from Portmahomack, Ross-shire, Scotland on 15 January 2018 at 23:44 UT.

passes below a bright cluster of stars, the **Pleiades,** or Seven Sisters. Even the smallest pair of binoculars reveals this cluster to be a beautiful group of bluish-white stars. The two most conspicuous of the other stars in Taurus lie directly above Orion, and form an elongated triangle with Aldebaran. The northernmost, β Tauri, was once considered to be part of the constellation of Auriga.

The Moon's phases for January

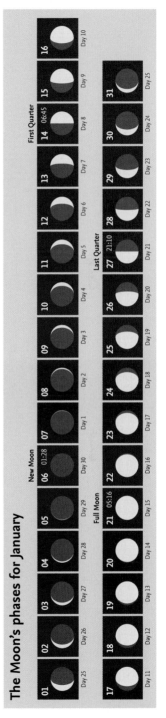

New Moon

Full Moon

Last Quarter

First Quarter

Day						
01 Day 25	02 Day 26	03 Day 27	04 Day 28	05 Day 29	06 01:28 Day 30	07 Day 1
08 Day 2	09 Day 3	10 Day 4	11 Day 5	12 Day 6	13 Day 7	14 06:45 Day 8
15 Day 9	16 Day 10					
17 Day 11	18 Day 12	19 Day 13	20 Day 14	21 05:16 Day 15	22 Day 16	23 Day 17
24 Day 18	25 Day 19	26 Day 20	27 21:10 Day 21	28 Day 22	29 Day 23	30 Day 24
31 Day 25						

January – Moon and Planets

The Earth

The Earth reaches perihelion (the closest point to the Sun in its annual orbit) on 3 January 2019, at 05:20 Universal Time. Its distance is then 0.9833 AU (147,099,586 km).

The Moon

On January 3 the Moon is close to Jupiter in Scorpius. On January 6 at New Moon there is a partial solar eclipse, visible from the region of the northwestern Pacific and northeastern Asia. On January 17, the Moon is close to **Aldebaran** in **Taurus**. A total lunar eclipse occurs at Full Moon on January 21, visible from a wide area of the Pacific, including Australia and eastern Asia. The Moon passes close to **Regulus** in **Leo** on January 23.

The planets

Mercury is too close to the Sun to be readily visible this month. **Venus** reaches greatest elongation west (47°) on January 6, when its magnitude is -4.6. **Mars** moves across the constellation of **Pisces**, fading from mag. 0.5 to 0.9 over the month. **Jupiter** is fairly bright (mag. -1.9 to -1.8) and in **Ophiuchus**, seen only in the early morning. **Saturn** is in **Sagittarius**, too near the Sun to be visible. **Uranus** (mag. 5.8) is on the border of **Pisces** and **Aries**. **Neptune** (mag. 7.9) is in **Aquarius**, where it remains throughout the year.

The path of the Sun and the planets along the ecliptic in January.

Calendar for January

*These objects are close together for an extended
period around this time.*

Morning 7:00

January 1–3 • *The Moon passes Venus, Antares and Jupiter, early in the morning.*

Evening 18:20

January 17 • *The Moon is close to Aldebaran, high in the southeast.*

Morning 7:00

January 22 • *Venus and Jupiter close together, in the company of Sabik and Antares.*

Morning 7:00

January 30 – February 1 • *The Moon with Antares, Jupiter, Sabik and Venus, in the morning sky.*

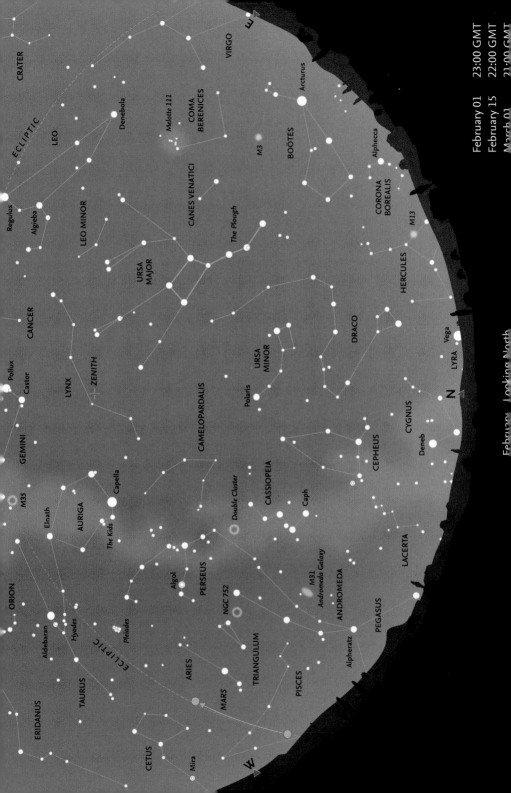

CRATER

ECLIPTIC

LEO

Denebola

Melotte 111

COMA
BERENICES

VIRGO

Arcturus

M3

BOÖTES

Regulus

Algieba

LEO MINOR

CANES VENATICI

The Plough

Alphecca

CORONA
BOREALIS

M13

HERCULES

CANCER

URSA
MAJOR

Pollux

Castor

LYNX

ZENITH

URSA
MINOR

Polaris

DRACO

Vega

LYRA

GEMINI

CAMELOPARDALIS

N

CYGNUS

Deneb

M35

Capella

AURIGA

Double Cluster

CASSIOPEIA

Caph

CEPHEUS

Elnath

The Kids

ORION

Algol

PERSEUS

NGC 752

Andromeda Galaxy

M31

ANDROMEDA

LACERTA

Aldebaran

Hyades

Pleiades

ECLIPTIC

ARIES

TRIANGULUM

PISCES

Alpheratz

PEGASUS

TAURUS

MARS

ERIDANUS

CETUS

Mira

W

February Looking North

February – Looking North

The months of January and February are probably the best time for seeing the section of the Milky Way that runs in the northern and western sky from **Cygnus**, low on the northern horizon, through **Cassiopeia**, **Perseus** and **Auriga** and then down through **Gemini** and **Orion**. Although not as readily visible as the denser star clouds of the summer Milky Way, on a clear night so many stars may be seen that even a distinctive constellation such as **Cassiopeia** is not immediately obvious.

The head of **Draco** is now higher in the sky and easier to recognize. **Deneb**, (α Cygni), the brightest star in **Cygnus**, may just be visible almost due north at midnight, early in the month, if the sky is very clear and the horizon clear of obstacles. **Vega** (α Lyrae) in **Lyra** is so low that it is difficult to see, but may become visible later in the night. The constellation of **Boötes** – sometimes described as shaped like a kite, an ice-cream cone, or the letter 'P' – with orange-tinted **Arcturus** (α Boötis), is beginning to clear the eastern horizon. Arcturus, at magnitude -0.05, is the brightest star in the northern hemisphere. The inconspicuous constellation of **Coma Berenices** is now well above the horizon in the east. The concentration of faint stars at the northwestern corner somewhat resembles a tiny, detached portion of the Milky Way. This is Melotte 111, an open star cluster (which is sometimes called the Coma Cluster, but must not be confused with the important Coma Cluster of galaxies, Abell 1656, mentioned on page 41).

On the other side of the sky, in the northwest, most of the constellation of **Andromeda** is still easily seen, although **Alpheratz** (α Andromedae), the star that forms the northeastern corner of the Great Square of Pegasus – even though it is actually part of Andromeda – is becoming close to the horizon and more difficult to detect. High overhead, at the zenith, try to make out the very faint constellation of

Lynx. It was introduced in 1687 by the famous astronomer Johannes Hevelius to fill the largely blank area between **Auriga**, **Gemini** and **Ursa Major**, and is reputed to be so named because one needed the eyes of a lynx to detect it.

A very large, and frequently ignored, open star cluster, Melotte 111, also known as the Coma Cluster, is readily visible in the eastern sky during February.

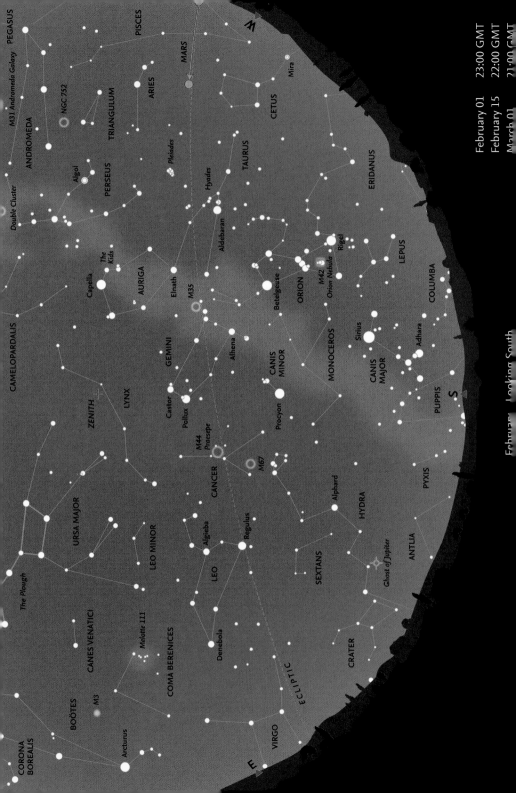

February 01   23:00 GMT
February 15   22:00 GMT
March 01   21:00 GMT

February Looking South

N

PEGASUS
PISCES
M31 Andromeda Galaxy
ANDROMEDA
NGC 752
TRIANGULUM
ARIES
PERSEUS
Algol
Double Cluster
CAMELOPARDALIS
The Kids
Capella
AURIGA
ZENITH
LYNX
URSA MAJOR
LEO MINOR
Melotte 111
CANES VENATICI
The Plough
M3
BOÖTES
CORONA
BOREALIS
Arcturus
COMA BERENICES
Denebola
LEO
Algieba
Regulus
CANCER
M44
Praesepe
M67
GEMINI
Castor
Pollux
M35
Elnath
TAURUS
Hyades
Pleiades
Aldebaran
Alhena
Betelgeuse
ORION
M42
Orion Nebula
Rigel
CETUS
Mira
MARS
ERIDANUS
LEPUS
COLUMBA
Adhara
CANIS
MAJOR
Sirius
MONOCEROS
CANIS
MINOR
Procyon
PUPPIS
S
PYXIS
HYDRA
Alphard
ANTLIA
Ghost of Jupiter
CRATER
SEXTANS
ECLIPTIC
VIRGO
E

February – Looking South

Apart from Orion, the most prominent constellation visible this month is *Gemini*, with its two lines of stars running southwest towards *Orion*. Many people have difficulty in remembering which is which of the two stars *Castor* and *Pollux*. Think of them in alphabetical order: Castor (α Geminorum), the fainter star, is closer to the North Celestial Pole. Pollux (β Geminorum) is the brighter of the two, but is farther away from the Pole. Castor is remarkable because it is actually a multiple system, consisting of no less than six individual stars.

Using Orion's belt as a guide, it points down to the southeast towards *Sirius*, the brightest star in the sky (at magnitude -1.4) in the constellation of *Canis Major*, the whole of which is now clear of the southern horizon. Forming an equilateral triangle with *Betelgeuse* in *Orion* and *Sirius* in Canis Major is *Procyon*, the brightest star in the small constellation of *Canis Minor*. Between Canis Major and Canis Minor is the faint constellation of *Monoceros*, which actually straddles the Milky Way, which, although faint, has many clusters in this area. Directly east of Procyon is the highly distinctive asterism of six stars that form the 'head' of *Hydra*, the largest

The constellation of Gemini. The two brightest stars are Castor and Pollux, visible on the left-hand side of the photograph.

of all 88 constellations, and which trails such a long way across the sky that it is only in mid-March around midnight that the whole constellation becomes visible.

The Moon's phases for February

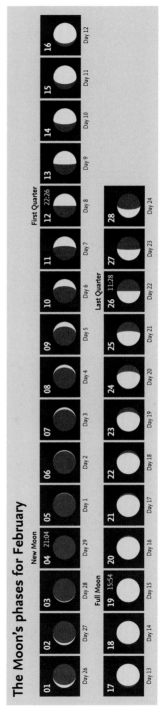

February – Moon and Planets

The Moon

The Moon is at its most distant apogee on February 5. It is 0.7°N of *Saturn* on February 2 and is 1.2°S of Venus later that day. On February 14, it is 1.7°N of *Aldebaran*, and five days later (February 19) is 2.4°N of *Regulus*. It is 2.5°N of *Jupiter* on February 27.

Occultations

Of the bright stars near the ecliptic that may be occulted by the Moon (*Aldebaran*, *Antares*, *Pollux*, *Regulus* and *Spica*), none are occulted in 2019. (There are occultations of *Saturn* in 2019, visible from the southern hemisphere, but not of any other planets.)

The planets

Mercury is lost in daylight, and although it reaches greatest eastern elongation on February 27, is too low to be readily visible. *Venus* is visible in the morning sky, rapidly moving closer to the Sun and fading slightly from mag. -4.3 to -4.1 over the month. *Mars* (at mag. 0.9 to 1.2) moves from *Pisces* into *Aries* in mid-month. *Jupiter* (mag. -1.9 to -2.0) is slowly moving east in *Ophiuchus*. *Saturn*, in *Sagittarius*, is also moving very slowly eastwards at mag. 0.6. *Uranus* (mag. 5.8) is in *Aries*, just inside the border with *Pisces*. *Neptune* (mag. 7.9) remains in *Aquarius*.

The path of the Sun and the planets along the ecliptic in February.

Calendar for February

07:18	Saturn 0.7°S of Moon
21:27	Venus 1.2°N of Moon
21:04	New Moon
07:02	Mercury 6.1°N of Moon
09:29	Moon at apogee (farthest of year, 406,555 km)
12:37	Mercury 8.4°N of Moon
16:19	Mars 6.1°N of Moon
22:26	First Quarter
02:29	Aldebaran 1.7°S of Moon
03:05	Moon 0.6°S of Praesepe (Beehive cluster)
14:16	Venus 1.1°N of Saturn
09:03	Moon at perigee (closest of year, 356,761 km)
13:08	Regulus 2.4°S of Moon
15:54	Full Moon
11:28	Last Quarter
01:25	Mercury at greatest elongation (18.1°E, mag. –0.5)
14:16	Jupiter 2.5°S of Moon

After midnight 1:00

February 14 • Shortly after midnight the Moon and Aldebaran approach the horizon.

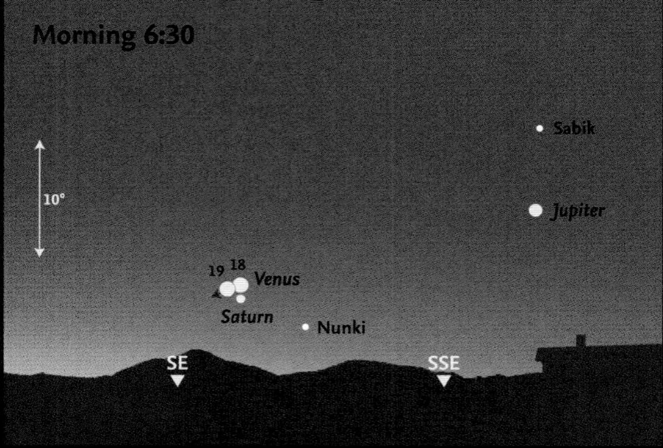

Morning 6:30

February 18–19 • Venus passes north of Saturn, low in the southeast. Jupiter is higher and farther south.

Evening 18:30

February 19 • The Moon is close to Regulus and Algieba when they rise in the east.

Morning 6:00

February 26–28 • The Moon passes Antares, Sabik and Jupiter. Nunki is closer to the horizon and may not be easy to detect.

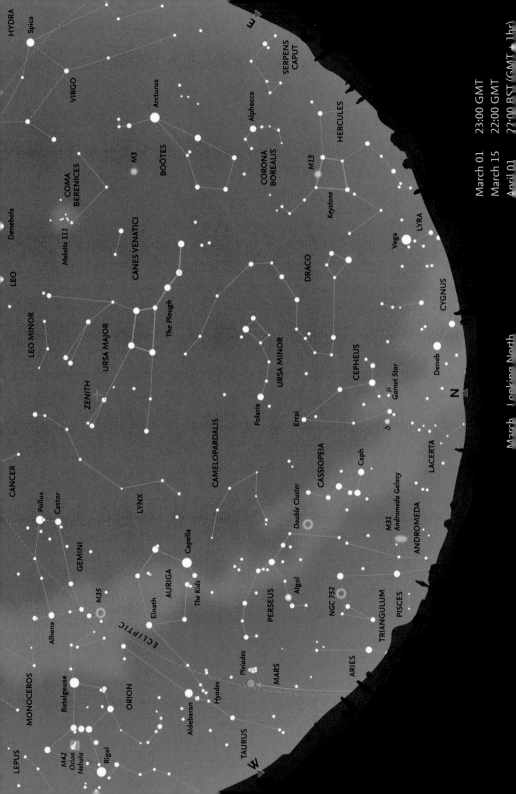

March 01   23:00 GMT
March 15   22:00 GMT
April 01   22:00 BST (GMT +1hr)

March Looking North

March – Looking North

In March, the Sun crosses the celestial equator on Wednesday, March 20, at the vernal equinox, when day and night are of almost exactly equal length, and the northern season of spring is considered to have begun. (The hours of daylight and darkness change most rapidly around the equinoxes, in March and September.) It is also in March that Summer Time begins in Europe (on Sunday, March 31) so the charts show the appearance at 23:00 GMT for March 1 and 22:00 BST for April 1. (In North America, Daylight Saving Time is introduced three weeks earlier, on Sunday, March 10.)

Early in the month, the constellation of **Cepheus** lies almost due north, with the distinctive 'W' of **Cassiopeia** to its west. Cepheus lies across the border of the Milky Way and is often described as like the gable-end of a house or a church tower and steeple. Despite the large number of stars revealed at the base of the constellation by binoculars, one star stands out because of its deep red colour. This is Mu (μ) Cephei, also known as the Garnet Star, because of its striking colour. It is a truly gigantic star, a red supergiant, and one of the largest stars known. It is about 2400 times the diameter of the Sun, and if placed in the Solar System would extend beyond the orbit of Saturn. (Betelgeuse, in Orion, is also a red supergiant, but it is 'only' about 500 times the diameter of the Sun.)

Another famous, and very important star in Cepheus is **δ Cephei**, which is the prototype for the class of variable stars known as Cepheids. These giant stars show a regular variation in their luminosity, and there is a direct relationship between the period of the changes in magnitude and the stars' actual luminosity. From a knowledge of the period of any Cepheid, its actual luminosity – known as its absolute magnitude – may be derived. A comparison of its apparent magnitude on the sky and its absolute magnitude enables the star's exact distance to be

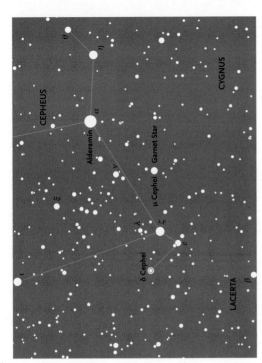

A finder chart for δ Cephei and μ Cephei (the Garnet Star). All stars brighter than magnitude 7.5 are shown.

determined. Once the distances to the first Cepheid variables had been established, examples in more distant galaxies provided information about the scale of the universe. Cepheid variables are the first 'rung' in the cosmic distance ladder. Both important stars are shown on the accompanying chart.

Below Cepheus to the east (to the right), it may be possible to catch a glimpse of **Deneb** (α Cygni), just above the horizon. Slightly farther round towards the northeast, **Vega** (α Lyrae) is marginally higher in the sky. From southern Britain, Deneb is just far enough north to be circumpolar (although difficult to see in January and February because it is so low on the northern horizon). Vega, by contrast, farther south, is completely hidden during the depths of winter.

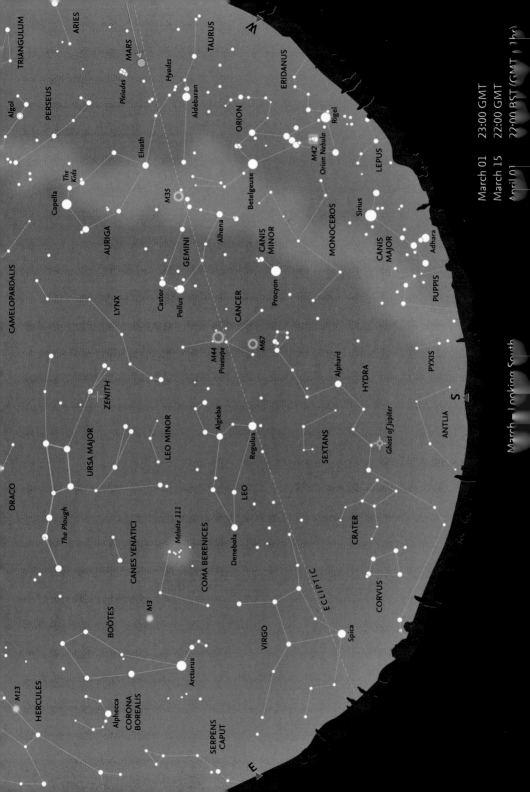

March 01   23:00 GMT
March 15   22:00 GMT
April 01   22:00 BST (GMT + 1hr)

March - Looking South

March – Looking South

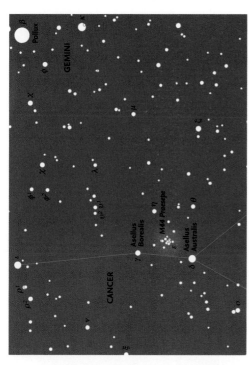

Due south at 22:00 at the beginning of the month, lying between the constellations of **Gemini** in the west and **Leo** in the east, and fairly high in the sky above the head of Hydra, is the faint, and rather undistinguished zodiacal constellation of **Cancer**. Rather like the triskelion, the symbol for the Isle of Man, it has three 'legs' radiating from the centre, where there is an open cluster, M44 or **Praesepe** ('the Manger' but also known as 'the Beehive'). On a clear night this cluster, known since antiquity, is just a hazy spot to the naked eye, but appears in binoculars as a group of dozens of individual stars.

Also prominent in March is the constellation of **Leo**, with the 'backward question mark' (or 'Sickle') of bright stars forming the head of the mythological lion. **Regulus** (α Leonis) – the 'dot' of the 'question mark' or the handle of the sickle and the brightest star in Leo – lies very close to the ecliptic and is one of the few first-magnitude stars that may be occulted by the Moon. However, there are no occultations of Regulus in 2019, nor of any of the other four bright stars near the ecliptic.

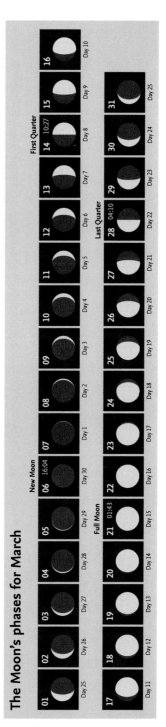

A finder chart for the open cluster M44 in Cancer. To the ancient Greeks and Romans, the two stars Asellus Borealis and Asellus Australis represented two donkeys feeding from Praesepe ('the Manger'). All stars brighter than magnitude 7.5 are shown.

The Moon's phases for March

					New Moon	
01	02	03	04	05	06 16:04	07
Day 25	Day 26	Day 27	Day 28	Day 29	Day 30	Day 1

						First Quarter	
08	09	10	11	12	13	14 10:27	15	16
Day 2	Day 3	Day 4	Day 5	Day 6	Day 7	Day 8	Day 9	Day 10

17	18	19	20	21 01:43	22	23
Day 11	Day 12	Day 13	Day 14	Day 15 Full Moon	Day 16	Day 17

24	25	26	27	28 04:10	29	30	31
Day 18	Day 19	Day 20	Day 21	Day 22 Last Quarter	Day 23	Day 24	Day 25

March – Moon and Planets

The Moon

The Moon (in *Sagittarius*) passes close to *Venus* (just within *Capricornus*) in the morning sky on March 2. (The two bodies are closest later in the evening in Capricornus, when below the horizon.) At New, on March 6, the Moon is in *Aquarius* (as is the Sun). It is north of *Aldebaran* in *Taurus* on March 11 and *Regulus* in *Leo* on March 19. It is in *Virgo*, lying almost half-way between *Spica* and Regulus, at Full Moon on March 21. It passes close to *Jupiter* on March 27 and *Saturn* on both March 1 and March 29.

The planets

Mercury is lost in daylight (it passes inferior conjunction on March 15). *Venus* at mag. -4.0 is visible in the morning sky, crossing *Capricornus* into *Aquarius*, close to the Sun at the end of the month. *Mars* is an evening object, initially mag. 1.2 and in *Aries*, moves into *Taurus* and fades slightly (to mag. 1.4) by the end of the month. *Jupiter* (mag. -2.0 to -2.2) is moving slowly eastwards in Ophiuchus. *Saturn* (mag. 0.6) is also moving slowly eastwards in *Sagittarius*. *Uranus* is mag. 5.9 in *Aries*, and *Neptune* (mag. 7.9) remains in *Aquarius*.

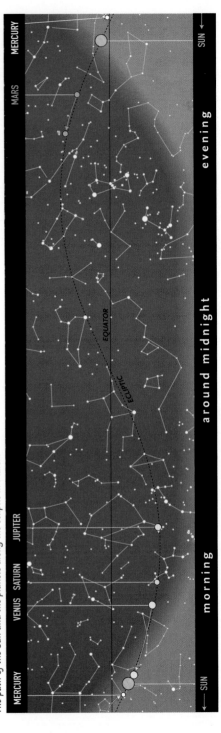

The path of the Sun and the planets along the ecliptic in March.

Morning 6:30

Venus
Saturn
Nunki
1
2
3
SSE
10°

March 1–3 • The Moon passes Nunki, Saturn and Venus, shortly before sunrise.

Evening 22:00

WNW
Pleiades
Mars
Aldebaran
13
12
11
E
10°

March 11–13 • The Moon in the western sky, with Mars, the Pleiades and Aldebaran.

Early morning 3:30

Sabik
Antares
Jupiter
26
27
S
SSE
10°

March 26–27 • Early in the morning the Moon passes Antares, Sabik and Jupiter.

Morning 5:00

Jupiter
Saturn
Nunki
28
29
S
SSE
10°

March 28–29 • The Moon in the company of Nunki and Saturn. Jupiter is almost due south at this time

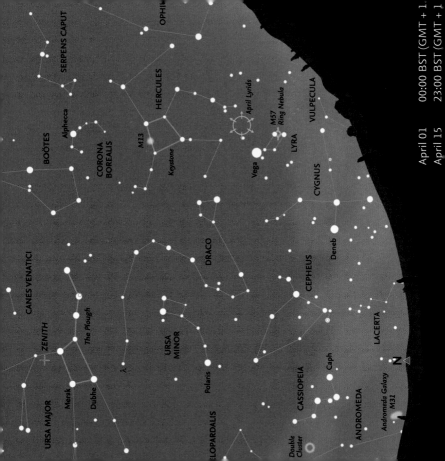

SERPENS CAPUT

OPHI

HERCULES

April Lyrids

M57
Ring Nebula

Alphecca

M13

VULPECULA

BOÖTES

CORONA
BOREALIS

Keystone

LYRA

Vega

CYGNUS

CANES VENATICI

DRACO

Deneb

ZENITH

The Plough

CEPHEUS

Merak

URSA
MINOR

LACERTA

λ

Dubhe

Polaris

N

URSA MAJOR

CASSIOPEIA

Caph

ELOPARDALIS

ANDROMEDA

Andromeda Galaxy
M31

Double
Cluster

April 01   00:00 BST (GMT + 1
April 15   23:00 BST (GMT + 1

April – Looking North

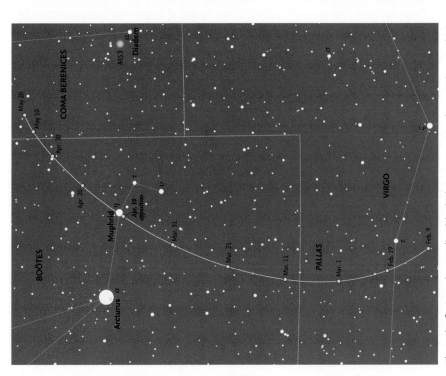

The path of minor planet Pallas (2) which is at opposition (mag. 7.9) on April 10. Background stars are shown down to magnitude 8.5.

Cygnus and the brighter regions of the Milky Way are now becoming visible, running more-or-less parallel with the horizon in the early part of the night. Rising in the northeast is the small constellation of *Lyra* and the distinctive 'Keystone' of **Hercules** above it. This asterism is very useful for locating the bright globular cluster M13 (see map on page 53), which lies on one side of the quadrilateral. The winding constellation of **Draco** weaves its way from the quadrilateral of stars that marks its 'head', on the border with Hercules, to end at λ Draconis between **Polaris** (α Ursae Minoris) and the 'Pointers', **Dubhe** and **Merak** (α and β Ursae Majoris, respectively). *Ursa Major* is 'upside down' high overhead, near the zenith. The constellation of **Gemini** stands almost vertically in the west. **Auriga** is still clearly seen in the northwest, but, by the end of the month, the southern portion of **Perseus** is starting to dip below the northern horizon. The very faint constellation of **Camelopardalis** lies in the northwest between Polaris and the constellations of Auriga and Perseus.

Meteors

A moderate meteor shower, the **Lyrids**, peaks on April 21–22. Although the hourly rate is not very high (about 18 meteors per hour), the meteors are fast and some leave persistent trains. This year the maximum occurs shortly after the Full Moon, meaning conditions are not ideal for seeing the fainter meteors. The parent object is the non-periodic comet C/1861 G1 (Thatcher). Another, stronger shower, the **Eta Aquariids**, begins to be active around April 19, and comes to maximum in May.

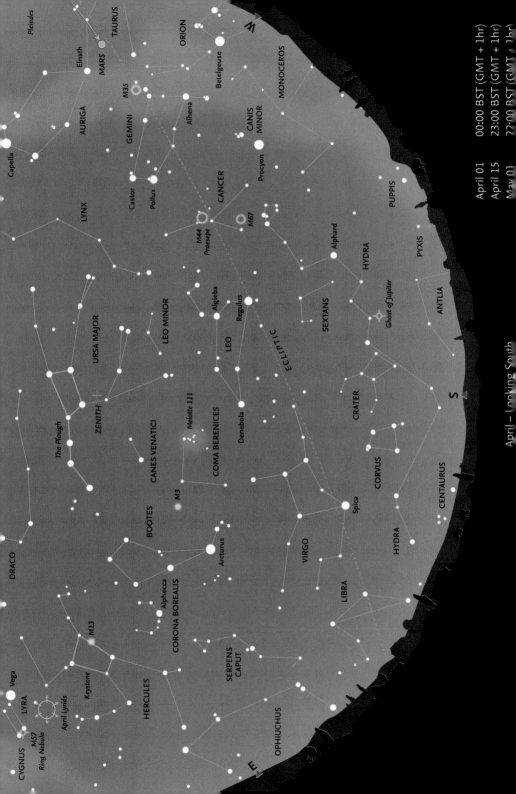

Pleiades

TAURUS

Elnath

MARS

M35

AURIGA

GEMINI

Castor

Pollux

Capella

LYNX

ORION

Betelgeuse

MONOCEROS

Alhena

CANIS
MINOR

Procyon

CANCER

M44
Praesepe

M67

Alphard

HYDRA

SEXTANS

PUPPIS

PYXIS

Ghost of Jupiter

ANTLIA

LEO MINOR

URSA MAJOR

Algieba

Regulus

LEO

ECLIPTIC

ZENITH

The Plough

Melotte 111

CANES VENATICI

COMA BERENICES

Denebola

CRATER

S

DRACO

BOÖTES

M3

Arcturus

Alphecca

CORONA BOREALIS

M13

CYGNUS

Vega

LYRA

M57
Ring Nebula

April Lyrids

Keystone

HERCULES

SERPENS
CAPUT

OPHIUCHUS

E

VIRGO

LIBRA

Spica

CORVUS

HYDRA

CENTAURUS

April 01   00:00 BST (GMT + 1hr)
April 15   23:00 BST (GMT + 1hr)
May 01   22:00 BST (GMT + 1hr)

April – Looking South

April – Looking South

Leo is the most prominent constellation in the southern sky in April, and vaguely looks like the creature after which it is named. **Gemini**, with **Castor** and **Pollux**, remains clearly visible in the west, and **Cancer** lies between the two constellations. To the east of Leo, the whole of **Virgo**, with **Spica** (α Virginis) its brightest star, is well clear of the horizon. Below Leo and Virgo, the complete length of **Hydra** is visible, running beneath both constellations, with **Alphard** (α Hydrae) halfway between Regulus and the southwestern horizon. Farther east, the two small constellations of **Crater** and the rather brighter **Corvus** lie between Hydra and Virgo.

Boötes and **Arcturus** are prominent in the eastern sky, together with the circlet of **Corona Borealis**, framed by Boötes and the neighbouring constellation of **Hercules**. Between Leo and Boötes lies the constellation of **Coma Berenices**, notable for being the location of the open cluster Melotte 111 (see page 27) and the Coma Cluster of galaxies (Abell 1656). There are about 1000 galaxies in this cluster, which is located near the North Galactic Pole, where we are looking out of the plane of the Galaxy and are thus able to see deep into space. Only about ten of the brightest galaxies in the Coma Cluster are visible with the largest amateur telescopes.

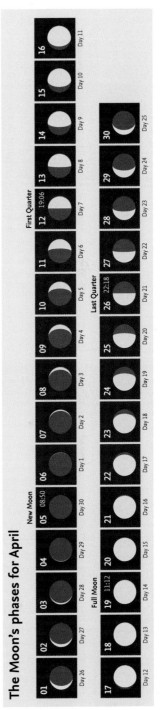

The distinctive constellation of Leo, with Regulus and 'The Sickle' on the west. Algieba (γ Leonis), north of Regulus, appearing double, is a multiple system of four stars.

The Moon's phases for April

					New Moon			
01	02	03	04	05 08:50	06	07	08	09
Day 26	Day 27	Day 28	Day 29	Day 30	Day 1	Day 2	Day 3	Day 4

Full Moon

10	11	12 19:06	13	14	15	16
Day 5	Day 6	Day 7	Day 8	Day 9	Day 10	Day 11

Last Quarter · First Quarter

17	18	19 11:12	20	21	22	23	24	25
Day 12	Day 13	Day 14	Day 15	Day 16	Day 17	Day 18	Day 19	Day 20

26 22:18	27	28	29	30
Day 21	Day 22	Day 23	Day 24	Day 25

April – Moon and Planets

The Moon

At New Moon on April 5, the Moon is in *Cetus,* below the Sun, which lies in *Pisces.* On April 9 it is visible in the western sky, north of *Aldebaran* in *Taurus.* On April 15 it passes *Regulus* in *Leo.* At Full Moon, on April 19, it is in *Virgo,* not far from *Spica.* On April 23, the Moon passes close to *Jupiter* and then *Saturn* on April 25. Both planets are fairly low in the south.

The planets

Mercury begins in *Aquarius* and rapidly moves to greatest western elongation on April 11. *Venus* also begins the month in Aquarius, at mag. -3.8, but then moves into the dawn twilight. Both planets are too low to be readily visible. *Mars,* in the evening sky, initially mag. 1.4, fades slightly to mag. 1.6 as it moves west in *Taurus* over the month. *Jupiter,* moving slowly in *Ophiuchus,* brightens slightly from mag. -2.2 to -2.4. and *Saturn* is in *Sagittarius* at mag. 0.6–0.5. Both planets are readily visible later in the night. *Uranus* is in *Aries* at mag. 5.9 and *Neptune* in *Aquarius* at mag. 8.0. Both planets are invisible in the daytime sky. Minor planet *Pallas* (2) comes to opposition in *Boötes* on April 10 at mag. 7.9 (see page 39).

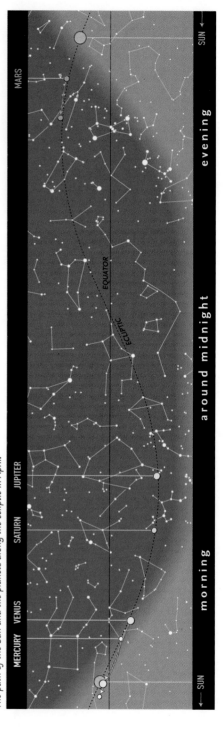

The path of the Sun and the planets along the ecliptic in April.

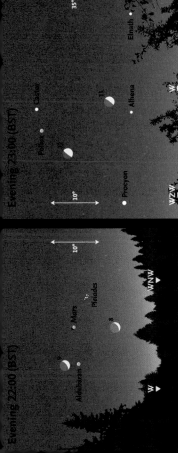

April 8–9 • The Moon with the Pleiades, Mars and Aldebaran, low in the western sky.

April 15 • The Moon with Regulus and

April 11–12 • High in the west the Moon is close to Alhena and, one day later, to Castor and Pollux.

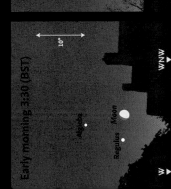

April 22–26 • The Moon passes Antares, Sabik, Jupiter, Nunki and

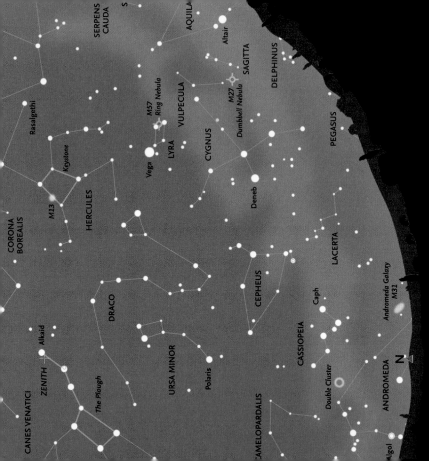

May – Looking North

Cassiopeia is now low over the northern horizon and, to its west, the southern portions of both **Perseus** and **Auriga** are becoming difficult to observe, although the **Double Cluster**, between Perseus and Cassiopeia is still clearly visible. Early in the night, the **Andromeda Galaxy** is too low to be visible with the extinction and other hindrances that occur close to the horizon. The constellations of **Lyra**, **Cepheus**, **Ursa Minor** and the whole of **Draco** are well placed in the sky. **Gemini**, with **Castor** and **Pollux**, is sinking towards the western horizon. **Capella** (α Aurigae) and the asterism of **The Kids** are still clear of the horizon.

In the east, two of the stars of the 'Summer Triangle', **Vega** (α Lyrae) and **Deneb** (α Cygni), are clearly visible, and the third star, **Altair** in **Aquila**, is beginning to climb above the horizon. The whole of **Cygnus** is now visible. The sprawling constellation of **Hercules** is high in the east and the brightest globular cluster in the northern hemisphere, M13, is visible to the naked eye on the western side of the asterism known as the **Keystone**.

Three faint constellations may be identified before the lighter nights of summer make them difficult objects. Below Cepheus, in the northeastern sky is the zig-zag constellation of **Lacerta**, while to the west, above Perseus and Auriga is **Camelopardalis** and, farther west, the line of faint stars forming **Lynx**.

Later in the night (and in the month) the westernmost stars of **Pegasus** begin to come into view, while the stars of **Andromeda** are skimming the northeastern horizon and the Andromeda Galaxy may become visible. High overhead, **Alkaid** (η Ursae Majoris), the last star in the 'tail' of the Great Bear, is close to the zenith, while the main body of the constellation has swung round into the western sky.

Meteors

The **Eta Aquariids** are one of the two meteor showers associated with Comet 1P/Halley (the other being the **Orionids**, in October). The Eta Aquariids are not particularly favourably placed for northern-hemisphere observers, because the radiant is near the celestial equator, near the 'Water Jar' in **Aquarius**, well below the horizon until late in the night (around dawn). However, meteors may still be seen in the eastern sky even when the radiant is below the horizon. There is a radiant map for the Eta Aquariids on page 16.

Their maximum in 2019, on May 6–7, occurs when the Moon is a narrow waxing crescent, just after New Moon, so conditions are particularly favourable. Maximum hourly rate is about 55 per hour and a large proportion (about 25 per cent) of the meteors leave persistent trains.

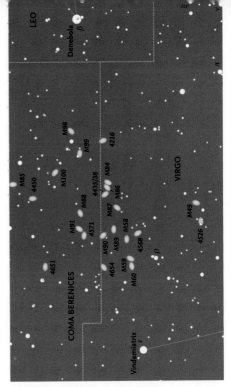

A finder chart for some of the brightest galaxies in the Virgo Cluster (see page 46). All stars brighter than magnitude 8.5 are shown.

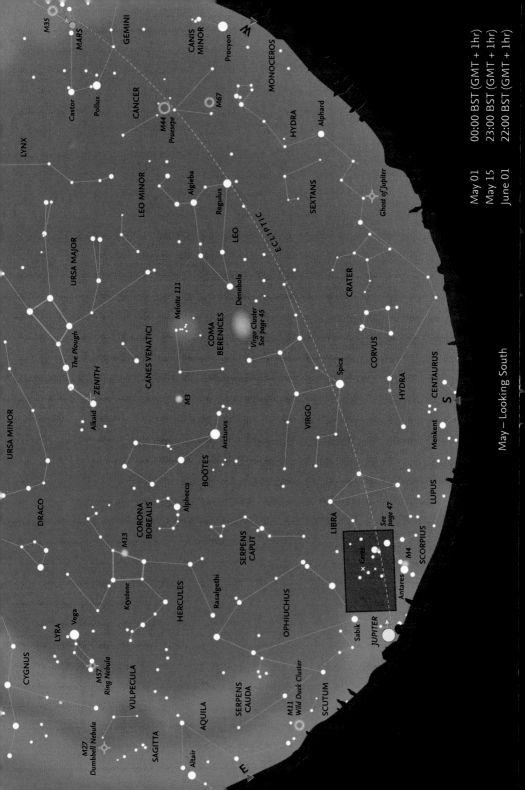

May 01   00:00 BST (GMT + 1hr)
May 15   23:00 BST (GMT + 1hr)
June 01   22:00 BST (GMT + 1hr)

May – Looking South

May – Looking South

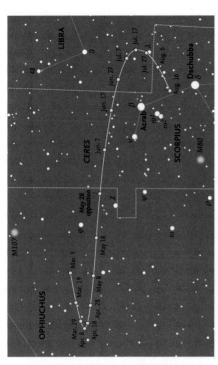

Early in the night, the constellation of **Virgo**, with **Spica** (α Virginis), lies due south, with **Leo** and both **Regulus** and **Denebola** (α and β Leonis, respectively) to its west still well clear of the horizon. Later in the night, the rather faint zodiacal constellation of **Libra** becomes visible and, to its east, the ruddy star **Antares** (α Scorpii) begins to climb up over the horizon.

Virgo contains the nearest large cluster of galaxies, which is the centre of the Local Supercluster, of which the Milky Way galaxy forms part. The Virgo Cluster contains some 2000 galaxies, the brightest of which are visible in amateur telescopes.

Arcturus in **Boötes** is high in the south, with the distinctive circlet of **Corona Borealis** clearly visible to its east. The brightest star (α Coronae Borealis) is known as **Alphecca**. The large constellation of **Ophiuchus** (which actually crosses the ecliptic, and is thus the 'thirteenth' zodiacal constellation) is climbing into the eastern sky. Before the constellation boundaries were formally adopted by the International Astronomical Union in 1930, the southern region of Ophiuchus was

The path of minor planet Ceres (1) which comes to opposition (mag. 7.0) on May 28. Background stars are shown down to magnitude 8.5.

regarded as forming part of the constellation of Scorpius, which had been part of the zodiac since antiquity.

The Moon's phases for May

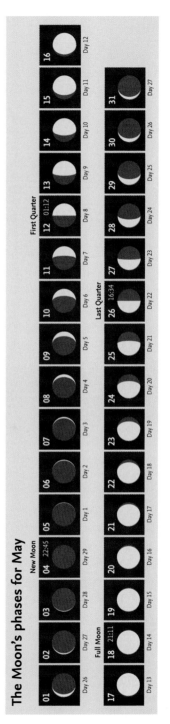

May – Moon and Planets

The Moon

New Moon occurs on May 4, when it is actually in the constellation of **Aries**, close to the border with **Cetus**. It is visible near to **Mars** on May 7 just before the planet sets in the west. It appears in **Leo**, not far from **Regulus** on May 12, at First Quarter. Full Moon is on May 18, in the constellation of **Libra**. It is close to **Jupiter** on May 20 and passes **Saturn** on May 22.

The planets

Mercury is invisible, close to the Sun, and is at superior conjunction on May 21. **Venus** is close to the horizon at dawn at the beginning of the month, but rapidly becomes invisible in daylight. **Mars** (mag. 1.6–1.7) may be glimpsed in **Taurus** in the evening twilight early in the month, but by the end of the month becomes too low as it moves into **Gemini**. **Jupiter** (mag. -2.5), and still in **Ophiuchus**, rises about 23:00 UT. **Saturn** (mag. 0.5), in **Sagittarius**, rises slightly later at about 01:00 and brightens slightly to mag. 0.3 over the month. Both planets are lost in twilight around 04:00 UT. **Uranus** is in **Aries** at mag. 5.9 and **Neptune** in **Aquarius** at mag. 7.9. The minor planet **Ceres** is at opposition (mag. 7.0) on May 28 in **Ophiuchus**, just inside the border with **Scorpius**.

The path of the Sun and the planets along the ecliptic in May.

Calendar for May

Evening 21:00 (BST)

May 6 • The Moon with Aldebaran, Mars and Elnath, shortly after sunset.

Evening 22:00 (BST)

May 7–9 • The Moon passes Mars, Elnath and Alhena. Procyon is farther southwest and Betelgeuse close to the horizon.

Evening 22:00 (BST)

May 12 • The Moon with Regulus and Algieba, high in the southwest.

Early morning 4:00 (BST)

May 19–23 • The Moon passes Antares, Sabik, Jupiter, Nunki and Saturn, low in the southern sky.

02	11:39	Venus 3.6°N of Moon
03	06:25	Mercury 2.9°N of Moon
04	22:45	New Moon
06–07		Eta Aquariid shower maximum
06	22:20	Aldebaran 2.3°S of Moon
07	23:35	Mars 3.2°N of Moon
10	03:56	Pollux 6.3°N of Moon
12	01:12	First Quarter
12	14:44	Regulus 3°S of Moon
13	21:53	Moon at perigee (369,009 km)
16	06:37	Spica 7.7°S of Moon
18	21:11	Full Moon
19	17:05	Antares 7.9°S of Moon
20	16:54	Jupiter 1.7°S of Moon
21	13:07	Mercury at superior conjunction
22	22:14	Saturn 0.5°N of Moon
26	13:27	Moon at apogee (404,138 km)
26	16:34	Last Quarter
28	22:36	Ceres at opposition (mag. 7.0)

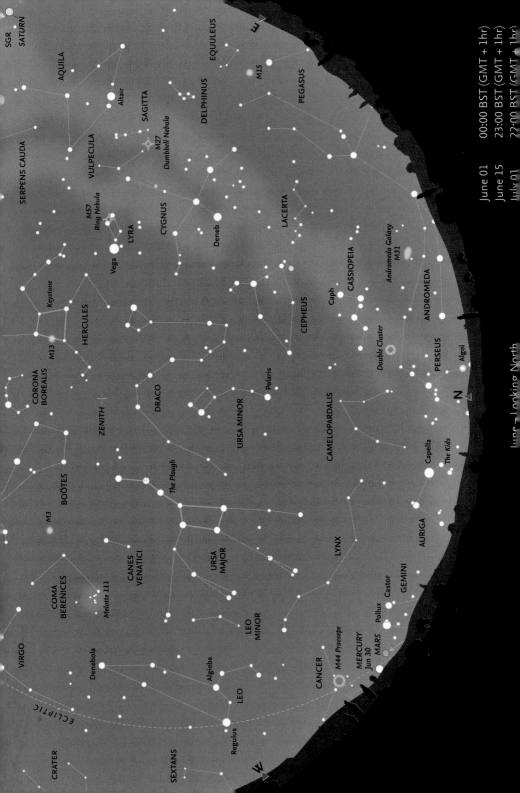

SGR
SATURN

AQUILA

EQUULEUS

Altair

SAGITTA

DELPHINUS

M15

PEGASUS

SERPENS CAUDA

VULPECULA

M27
Dumbbell Nebula

LACERTA

M57
Ring Nebula

CYGNUS

LYRA

Deneb

Vega

ANDROMEDA

Andromeda Galaxy
M31

Keystone

CASSIOPEIA

Caph

HERCULES

CEPHEUS

M13

Double Cluster

PERSEUS

Algol

CORONA
BOREALIS

DRACO

Polaris

CAMELOPARDALIS

Capella

The Kids

ZENITH

URSA MINOR

N

BOÖTES

The Plough

M3

AURIGA

CANES
VENATICI

URSA
MAJOR

LYNX

COMA
BERENICES

GEMINI

VIRGO

Melotte 111

LEO
MINOR

Castor

Pollux

Denebola

Algieba

CANCER

MARS

M44 Praesepe

MERCURY
Jun 30

ECLIPTIC

LEO

Regulus

CRATER

SEXTANS

W

June – Looking North

Around summer solstice (June 21) even in southern England and Ireland a form of twilight persists throughout the night. Farther north, in Scotland, the sky remains so light that most of the fainter stars and constellations are invisible. There, even brighter stars, such as the seven stars making up the well-known asterism known as the *Plough* in *Ursa Major* may be difficult to detect except around local midnight, 00:00 UT (01:00 BST).

But there is one compensation during these light nights: even southern observers may be lucky enough to witness a display of noctilucent clouds (NLC). These are highly distinctive clouds shining with an electric-blue tint, observed in the sky in the direction of the North Pole. They are the highest clouds in the atmosphere, occurring at altitudes of 80–85 km, far above all other clouds. They are only visible during summer nights, for about a month or six weeks on either side of the solstice, when observers are in darkness, but the clouds themselves remain illuminated by sunlight, reaching them from the Sun, itself hidden below the northern horizon.

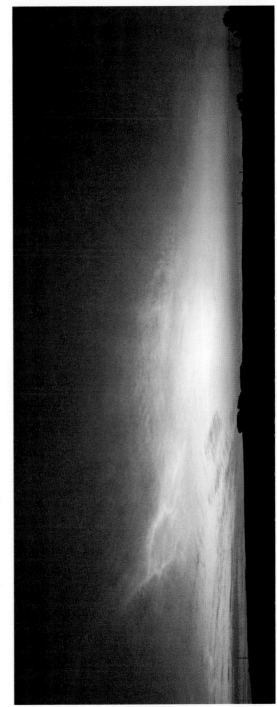

Noctilucent clouds, photographed on the night of 5–6 July 2016, from Portmahomack, Ross-shire, Scotland, by Denis Buczynski.

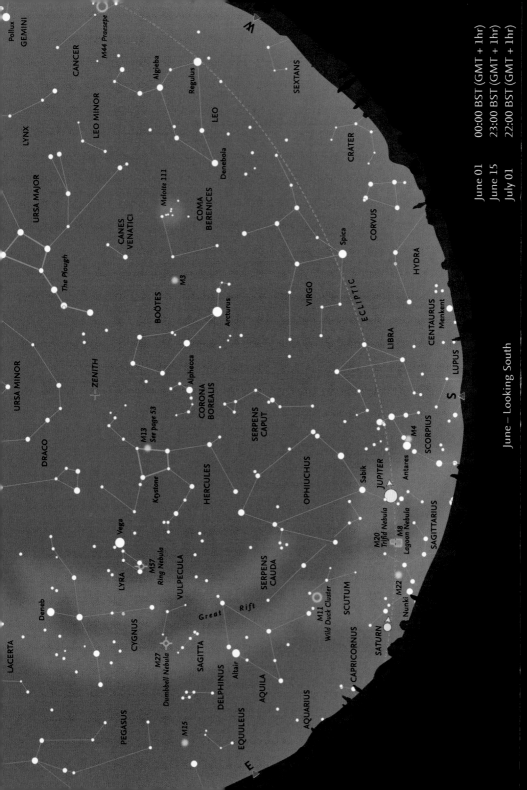

June – Looking South

June 01   00:00 BST (GMT + 1hr)
June 15   23:00 BST (GMT + 1hr)
July 01   22:00 BST (GMT + 1hr)

June – Looking South

Although the persistent twilight makes observing even the southern sky difficult, the rather undistinguished constellation of *Libra* lies almost due south. The red supergiant star *Antares* – the name means the 'Rival of Mars' – in *Scorpius* is visible slightly to the east of the meridian, but the 'tail' or 'sting' remains below the horizon. Higher in the sky is the large constellation of *Ophiuchus* (the 'Serpent Bearer'), lying between the two halves of the constellation of *Serpens*: *Serpens Caput* ('Head of the Serpent') to the west and *Serpens Cauda* ('Tail of the Serpent') to the east. (Serpens is the only constellation to be divided into two distinct parts.) The ecliptic runs across Ophiuchus, and the Sun actually spends far more time in the constellation than it does in the 'classical' zodiacal constellation of Scorpius, a small area of which lies between Libra and Ophiuchus.

Higher in the southern sky, the three constellations of *Boötes*, *Corona Borealis* and *Hercules* are now better placed for observation than at any other time of the year. This is an ideal time to observe the fine globular cluster of M13 in Hercules.

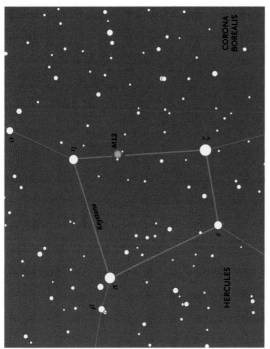

Finder chart for M13, the finest globular cluster in the northern sky. All stars down to magnitude 7.5 are shown.

The Moon's phases for June

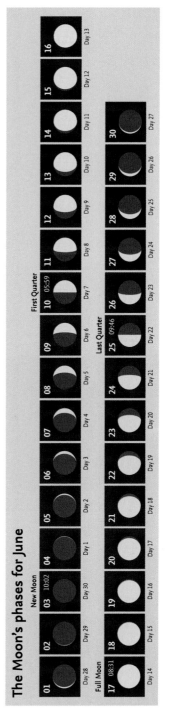

June – Moon and Planets

The Moon

New Moon is on June 3 when it is close to *Aldebaran* in *Taurus*, but the star is invisible in morning twilight. Later in the month the Moon is again near Aldebaran on June 30, but the star is now in daylight. It passes close to *Mars* on June 5, and both bodies may be glimpsed later in *Gemini* just before they set in the west. It passes *Spica* in *Virgo* in daylight on June 12 and again both bodies may be seen later in the day as they begin to set in the southwest. It is in *Ophiuchus*, above *Antares* (in *Scorpius*) on June 16 and passes close to *Jupiter* later that day. Full Moon is on June 17 (in *Ophiuchus*, near Jupiter). It passes very close to *Saturn* (in *Sagittarius*) on June 19, but the event occurs in the southwest in early evening twilight.

The planets

Mercury is initially in the daytime sky, but rapidly moves to greatest eastern elongation on June 23, when it is actually close to *Mars*, but too low to be visible after sunset. It may be possible to glimpse *Venus* (at mag. -3.9) early in the month, very low in morning twilight, but it soon moves into daylight. *Mars* is in *Gemini*, but invisible in daylight, although moving into the western evening sky. *Jupiter* is in *Ophiuchus* at mag. -2.6 and comes to opposition on June 10, *Saturn* (mag. 0.3–0.1) is in *Sagittarius*. *Uranus* remains in Aries at mag. 5.9 and *Neptune* in *Aquarius* at mag. 7.9.

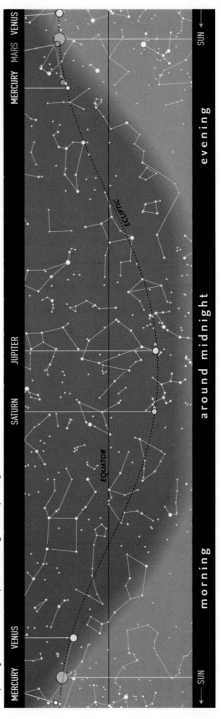

The path of the Sun and the planets along the ecliptic in June.

Calendar for June

01	18:14	Venus 3.2°N of Moon
03	06:12	Aldebaran 2.3°S of Moon
03	10:02	New Moon
04	15:41	Mercury 3.7°N of Moon
05	15:05	Mars 1.6°N of Moon
06	10:07	Pollux 6.2°N of Moon
07	23:15	Moon at perigee (368,504 km)
08	20:01	Regulus 3.2°S of Moon
10	05:59	First Quarter
10	15:28	Jupiter at opposition (mag. -2.6)
12	12:49	Spica 7.8°S of Moon
16	01:02	Antares 8°S of Moon
16	18:50	Jupiter 2°S of Moon
17	08:31	Full Moon
17	21:00 *	Venus 4.8°N of Aldebaran
18	15:00 *	Mars 0.2°S of Mercury
19	03:46	Saturn 0.4°N of Moon
21	15:54	Summer solstice
23	07:00 *	Mars 5.6°S of Pollux
23	07:50	Moon at apogee (404,548 km)
23	23:16	Mercury at greatest elongation (25.2°E, mag. 0.4)
25	09:46	Last Quarter
30	15:34	Aldebaran 2.3°S of Moon

* These objects are close together for an extended period around this time.

Evening 22:00 (BST)

June 4–6 • The narrow crescent Moon near Mercury, Mars and Pollux, shortly after sunset and low in the western sky.

Evening 22:30 (BST)

June 8 • The Moon is lining up with Regulus and Algieba.

Early morning 2:00 (BST)

June 16–19 • The Moon passes Antares, Sabik, Jupiter, Nunki and Saturn, low in the south.

Evening 22:15 (BST)

June 18 • Mercury and Mars, close together after sunset. Castor and Pollux are nearby.

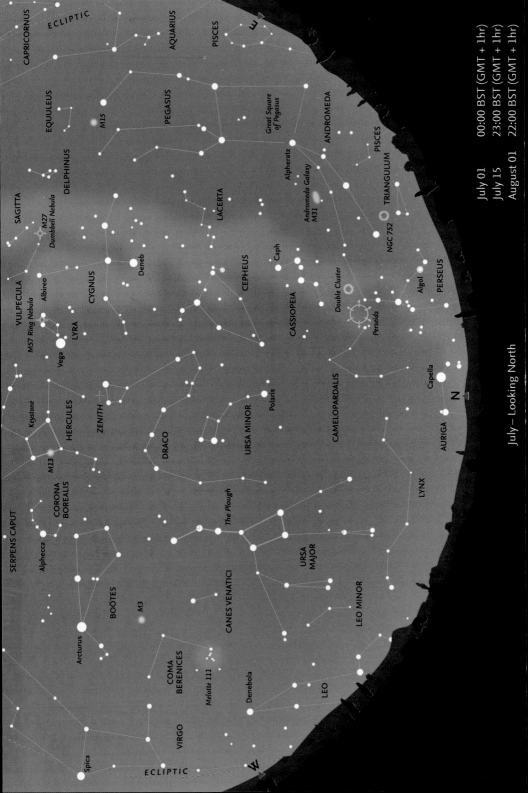

ECLIPTIC
CAPRICORNUS
AQUARIUS
PISCES
E

EQUULEUS
M15
PEGASUS
Great Square
of Pegasus
ANDROMEDA
PISCES
Alpheratz
Andromeda Galaxy
M31
TRIANGULUM
NGC 752
PERSEUS
Algol

DELPHINUS
SAGITTA
M27
Dumbbell Nebula
LACERTA
Caph
Double Cluster
Perseids

VULPECULA
Albireo
M57 Ring Nebula
LYRA
Vega
CYGNUS
Deneb
CEPHEUS
CASSIOPEIA
CAMELOPARDALIS
Capella
N

HERCULES
Keystone
ZENITH
DRACO
URSA MINOR
Polaris
AURIGA

M13
CORONA
BOREALIS
Alphecca
SERPENS CAPUT
The Plough
URSA
MAJOR
LYNX

BOÖTES
M3
Arcturus
CANES VENATICI
LEO MINOR

COMA
BERENICES
Melotte 111
Denebola
LEO

VIRGO
Spica
ECLIPTIC
W

July – Looking North

As in June, light nights and the chance of observing noctilucent clouds persist throughout July, but later in the month (and particularly after midnight) some of the major constellations begin to be more easily seen. **Capella**, the brightest star in **Auriga** (most of which is too low to be visible), is skimming the northern horizon. **Cassiopeia** is clearly visible in the northeast and **Perseus**, to its south, is beginning to climb clear of the horizon. The band of the Milky Way, from Perseus through Cassiopeia towards **Cygnus**, stretches up into the northeastern sky. If the sky is dark and clear, you may be able to make out the small, faint constellation of **Lacerta**, lying across the Milky Way between Cassiopeia and Cygnus. In the east, the stars of **Pegasus** are now well clear of the horizon, with the main line of stars forming **Andromeda** roughly parallel to the horizon in the northeast. **Alpheratz** (α Andromedae) is actually the star at the northeastern corner of the **Great Square of Pegasus. Cepheus** and **Ursa Major** are on opposite sides of **Polaris** and **Ursa Minor,** in the east and west, respectively. The head of **Draco** is very close to the zenith so the whole of this winding constellation is readily seen.

Meteors

July brings increasing meteor activity, mainly because there are several minor radiants active in the constellations of **Capricornus** and **Aquarius.** Because of their location, however, observing conditions are not particularly favourable for northern-hemisphere observers, although the first shower, the **Alpha Capricornids,** active from July 11 to August 10 (peaking July 26 to August 1), does often produce very bright fireballs. The maximum rate, however, is only about 5 per hour. The parent body is Comet 169P/NEAT. The most prominent shower is probably that of the **Delta Aquariids,** which are active from around July 21 to August 23, with a peak on July 29–30, although even then the hourly rate is unlikely to reach 20 meteors per hour. In this case, the

Cygnus, sometimes known as the 'Northern Cross', depicts a swan flying down the Milky Way towards Sagittarius. The brightest star, Deneb (α Cygni), represents the tail, and Albireo (β Cygni) marks the position of the head, and may be found at bottom-right of this image.

parent body is possibly Comet 96P/Machholz. This year both shower maxima occur when the Moon is a waning crescent, so observing conditions are reasonably favourable. A chart showing the **Delta Aquariid** radiant is shown on page 16. The **Perseids** begin on July 13 and peak on August 11–12.

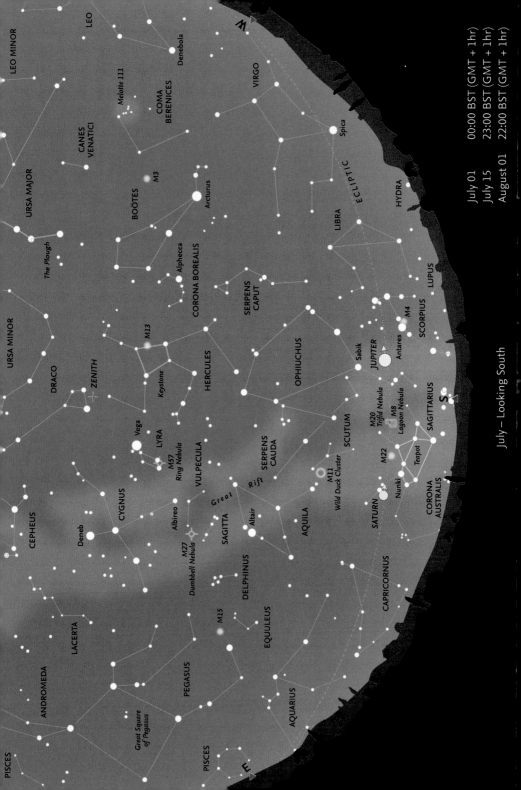

LEO MINOR

LEO

URSA MAJOR

CANES VENATICI

Melotte 111

COMA BERENICES

Denebola

VIRGO

Spica

The Plough

M3

BOÖTES

Arcturus

Alphecca

CORONA BOREALIS

ECLIPTIC

LIBRA

HYDRA

URSA MINOR

SERPENS CAPUT

M13

LUPUS

M4

SCORPIUS

DRACO

ZENITH

Keystone

HERCULES

OPHIUCHUS

Sabik

JUPITER

Antares

Vega

LYRA

M57
Ring Nebula

VULPECULA

SERPENS CAUDA

SCUTUM

M20
Trifid Nebula

M8
Lagoon Nebula

SAGITTARIUS

CYGNUS

Deneb

Great

Rift

M11
Wild Duck Cluster

M22

Teapot

CEPHEUS

Albireo

SAGITTA

Altair

AQUILA

SATURN

Nunki

CORONA AUSTRALIS

M27
Dumbbell Nebula

DELPHINUS

LACERTA

EQUULEUS

M15

CAPRICORNUS

ANDROMEDA

PEGASUS

*Great Square
of Pegasus*

AQUARIUS

PISCES

PISCES

July 01   00:00 BST (GMT + 1hr)
July 15   23:00 BST (GMT + 1hr)
August 01   22:00 BST (GMT + 1hr)

July – Looking South

July – Looking South

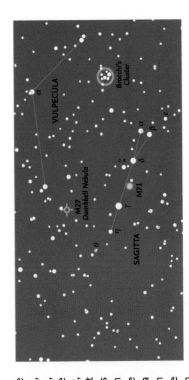

Although part of the constellation remains hidden, this is perhaps the best time of year to see *Scorpius*, with deep red *Antares* (α Scorpii), glowing just above the southern horizon. At around midnight (UT), 01:00 BST, part of *Sagittarius*, with the distinctive asterism of the 'Teapot', and the dense star clouds of the centre of the Milky Way, are just visible in the south. The *Great Rift* – actually dust clouds that hide the more distant stars – runs down the Milky Way from Cygnus towards Sagittarius. Towards its northen end is the small constellation of *Sagitta* and the planetary nebula *M27* (the Dumbbell Nebula). The sprawling constellation of *Ophiuchus* lies close to the meridian for a large part of the month, separating the two halves of the constellation of *Serpens*. The western half is called *Serpens Caput* (Head of the Serpent) and the eastern part *Serpens Cauda* (Tail of the Serpent). In the east, the bright *Summer Triangle*, consisting of *Vega* in *Lyra*, *Deneb* in *Cygnus* and *Altair* in *Aquila*, begins to dominate the southern sky, as it will throughout August and into September. The small constellation of Lyra, with Vega and a distinctive quadrilateral of stars to its east and south, lies not far south of the zenith.

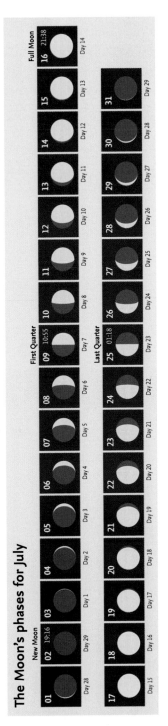

A finder chart for M27, the Dumbbell Nebula, a relatively bright (magnitude 8) planetary nebula – a shell of material ejected in the late stages of a star's lifetime – in the constellation of Vulpecula. All stars brighter than magnitude 7.5 are shown.

The Moon's phases for July

New Moon

01	02 19:16	03	04	05	06	07	08
Day 28	Day 29	Day 1	Day 2	Day 3	Day 4	Day 5	Day 6

First Quarter

09 10:55	10	11	12	13	14	15	16 21:38
Day 7	Day 8	Day 9	Day 10	Day 11	Day 12	Day 13	Day 14

Full Moon

17	18	19	20	21	22	23	24
Day 15	Day 16	Day 17	Day 18	Day 19	Day 20	Day 21	Day 22

Last Quarter

25 01:18	26	27	28	29	30	31	
Day 23	Day 24	Day 25	Day 26	Day 27	Day 28	Day 29	

July – Moon and Planets

The Moon

New Moon occurs on July 2, when the Moon (and Sun) are in **Gemini**. That day there is a total solar eclipse, visible from the South Pacific and the southern tip of South America. On July 4, the Moon is very close to **Mars** in **Cancer**, but both bodies are lost in morning twilight. It passes **Regulus** in **Leo** on July 6, but the constellation is just visible early in the night as it sets in the west. On July 9, the Moon is close to **Spica** in **Virgo**, and again the constellation is visible only later as it sets. On July 13 the Moon is close to **Antares** in **Scorpius** and then passes **Jupiter** in **Ophiuchus**. On July 16 it passes very close to **Saturn**, but Full Moon occurs later in the day (when both bodies rise in the east), so Saturn will be difficult to see. That day there is a partial lunar eclipse, visible from Africa and the Middle East. On July 28, the Moon is a waning crescent close to **Aldebaran**, in **Taurus**, visible as the objects rise in the east.

The planets

Mercury remains invisible in daylight and passes inferior conjunction (between Earth and the Sun) on July 21. **Venus** is also too close to the Sun to be detected. **Mars** remains close to the Sun in **Cancer**. **Jupiter** (mag. -2.6 to -2.4) is still retrograding slowly in **Ophiuchus** and **Saturn** (mag. 0.1) is doing the same in **Sagittarius**, reaching opposition on July 9. **Uranus** is still in **Aries** at mag. 5.9 and **Neptune** in **Aquarius** at mag. 7.9.

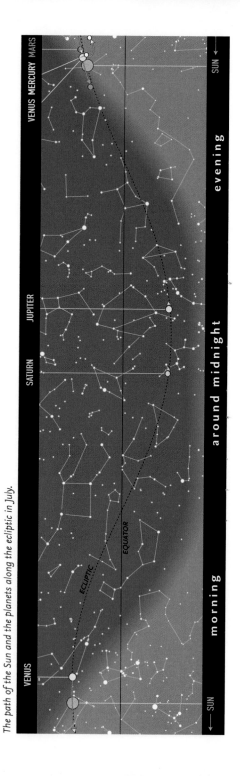

The path of the Sun and the planets along the ecliptic in July.

60

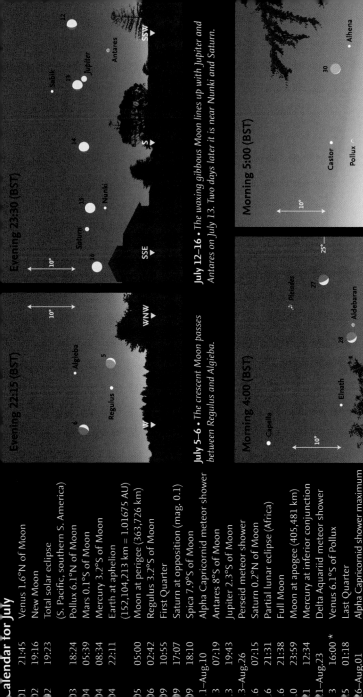

Calendar for July

01	21:45	Venus 1.6°N of Moon
02	19:16	New Moon
02	19:23	Total solar eclipse
		(S. Pacific, southern S. America)
03	18:24	Pollux 6.1°N of Moon
04	05:39	Mars 0.1°S of Moon
04	08:34	Mercury 3.2°S of Moon
04	22:11	Earth at aphelion
		(152,104,213 km = 1.01675 AU)
05	05:00	Moon at perigee (363,726 km)
06	02:42	Regulus 3.2°S of Moon
09	10:55	First Quarter
09	17:07	Saturn at opposition (mag. 0.1)
09	18:10	Spica 7.9°S of Moon
11–Aug.10		Alpha Capricornid meteor shower
13	07:19	Antares 8°S of Moon
13	19:43	Jupiter 2.3°S of Moon
13–Aug.26		Perseid meteor shower
16	07:15	Saturn 0.2°N of Moon
16	21:31	Partial lunar eclipse (Africa)
16	21:38	Full Moon
20	23:59	Moon at apogee (405,481 km)
21	12:34	Mercury at inferior conjunction
21–Aug.23		Delta Aquariid meteor shower
23	16:00 *	Venus 6.1°S of Pollux
25	01:18	Last Quarter
26–Aug.01		Alpha Capricornid shower maximum
28	01:16	Aldebaran 2.3°S of Moon
29–30		Delta Aquariid shower maximum
31	02:18	Mercury 4.5°S of Moon
31	04:29	Pollux 6.1°N of Moon
31	20:36	Venus 0.6°S of Moon

These objects are close together for an extended

Evening 22:15 (BST)

July 5–6 • The crescent Moon passes between Regulus and Algieba.

Evening 23:30 (BST)

July 12–16 • The waxing gibbous Moon lines up with Jupiter and Antares on July 13. Two days later it is near Nunki and Saturn.

Morning 4:00 (BST)

July 27–28 • The Moon with Aldebaran and the Pleiades, Elnath and Capella are nearby.

Morning 5:00 (BST)

July 30–31 • The Moon passes Alhena and Pollux shortly before sunrise. On July 31 it is probably lost in dawn

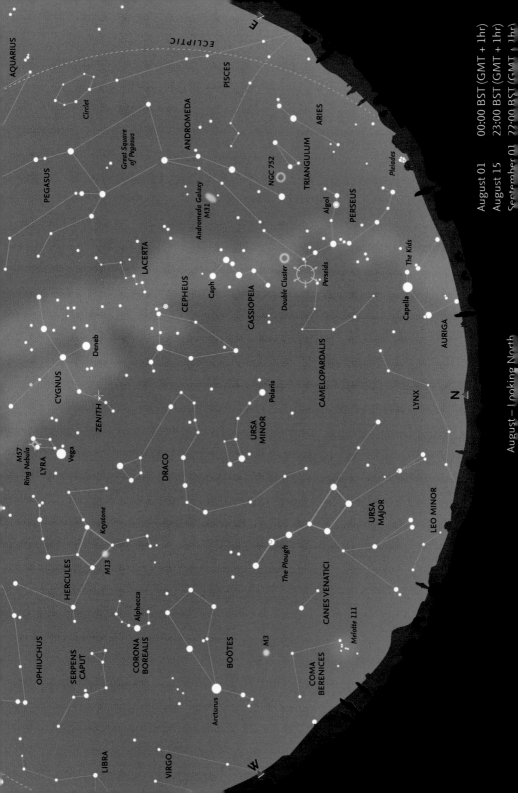

August – Looking North

August 01   00:00 BST (GMT + 1hr)
August 15   23:00 BST (GMT + 1hr)
September 01   22:00 BST (GMT + 1hr)

August – Looking North

A brilliant Perseid fireball, streaking alongside the Great Rift in the Milky Way, photographed in 2012 by Jens Hackmann from near Weikersheim in Germany. Four additional, fainter Perseids are also visible in the image.

Ursa Major is now the 'right way up' in the northwest, although some of the fainter stars in the south of the constellation are difficult to see. Beyond it, *Boötes* stands almost vertically in the west, but pale orange *Arcturus* is sinking towards the horizon. Higher in the sky, both *Corona Borealis* and *Hercules* are clearly visible.

In the northeast, *Capella* is clearly seen, but most of *Auriga* still remains below the horizon. Higher in the sky, *Perseus* is gradually coming into full view and, later in the night and later in the month, the beautiful *Pleiades* cluster rises above the northeastern horizon. Between Perseus and *Polaris* lies the faint and unremarkable constellation of *Camelopardalis*.

Higher still, both *Cassiopeia* and *Cepheus* are well placed for observation, despite the fact that Cassiopeia is completely immersed in the band of the Milky Way, as is the 'base' of Cepheus. *Pegasus* and *Andromeda* are now well above the eastern horizon and, below them, the constellation of *Pisces* is climbing into view. Two of the stars in the *Summer Triangle*, *Deneb* and *Vega*, are close to the zenith high overhead.

Meteors

August is the month when one of the best meteor showers of the year occurs: the *Perseids*. This is a long shower, generally beginning about July 13 and continuing until about August 26, with a maximum in 2019 on August 11–12, when the rate may reach as high as 100 meteors per hour (and on rare occasions, even higher). In 2019, maximum is just before Full Moon, so conditions are particularly unfavourable. The Perseids are debris from Comet 109P/Swift-Tuttle (the Great Comet of 1862). Perseid meteors are fast and many of the brighter ones leave persistent trains. Some bright fireballs also occur during the shower.

August – Looking South

August 01 00:00 BST (GMT + 1hr)
August 15 23:00 BST (GMT + 1hr)
September 01 22:00 BST (GMT + 1hr)

W

VIRGO

CANES VENATICI
Melotte 111
COMA BERENICES
M3
BOÖTES
Arcturus
Alphecca
CORONA BOREALIS
SERPENS CAPUT
LIBRA

The Plough
URSA MAJOR
M13
OPHIUCHUS
Sabik
SCORPIUS
JUPITER
Antares
M4

DRACO
HERCULES
Keystone
SERPENS CAUDA
M20
Trifid Nebula
M8
Lagoon Nebula
SAGITTARIUS

Vega
LYRA
M57
Ring Nebula
Great Rift
M11
Wild Duck Cluster
SCUTUM
SATURN
M22
Nunki
Teapot

CEPHEUS
ZENITH
Albireo
VULPECULA
CYGNUS
Deneb
M27
Dumbbell Nebula
SAGITTA
Altair
DELPHINUS
AQUILA
CORONA AUSTRALIS
S

Caph
CASSIOPEIA
LACERTA
M15
EQUULEUS
CAPRICORNUS
MICROSCOPIUM

PERSEUS
NGC 752
ANDROMEDA
Andromeda Galaxy
M31
PEGASUS
Great Square of Pegasus
Water Jar
AQUARIUS
Delta Aquarids
PISCIS AUSTRINUS

TRIANGULUM
ARIES
PISCES
Circlet
CETUS
ECLIPTIC
E

August – Looking South

The whole stretch of the summer Milky Way stretches across the sky in the south, from **Cygnus**, high in the sky near the zenith, past **Aquila**, with bright **Altair** (α Aquilae), to part of the constellation of **Sagittarius** close to the horizon, where the pattern of stars known as the 'Teapot' may be just visible. This area contains many nebulae and both open and globular clusters. Between **Albireo** (β Cygni) and Altair lie the two small constellations of **Vulpecula** and **Sagitta**, with the latter easier to distinguish (because of its shape) from the clouds of the Milky Way. Between Sagitta and **Pegasus** to the east lie the highly distinctive five stars that form the tiny constellation of **Delphinus** (again, one of the few constellations that actually bear some resemblance to the creatures after which they are named). Below Aquila, mainly in the star clouds of the Milky Way, lies **Scutum**, most famous for the bright open cluster, **M11** or the 'Wild Duck Cluster', readily visible in binoculars. To the southeast of Aquila lie the two zodiacal constellations of **Capricornus** and **Aquarius**.

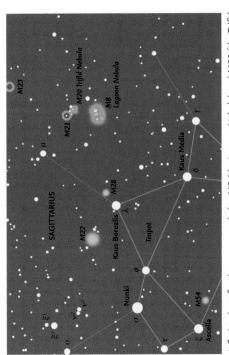

A finder chart for the gaseous nebulae M8 (the Lagoon Nebula) and M20 (the Trifid Nebula) and the globular cluster M22, all in Sagittarius. Clusters M21 & M23 (open) and M28 (globular) are faint. The chart shows all stars brighter than magnitude 7.5.

The Moon's phases for August

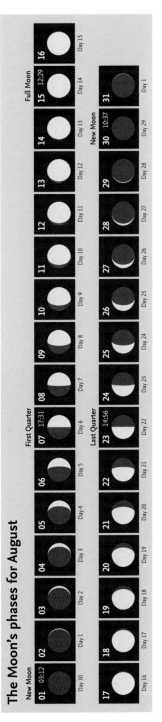

August – Moon and Planets

The Moon

New Moon occurs on August 1 in *Cancer*. A few hours later, it passes *Mars*, just inside *Leo* and the next day it is close to *Regulus*, but these events occur in daylight. On August 6, just before First Quarter, it passes *Spica* in *Virgo*, but the constellation is visible only as it sets in the west. On August 9, the Moon is close to both *Antares* in *Scorpius* and *Jupiter* (in *Ophiuchus*), visible just before they set. On August 12, the Moon is close to *Saturn* in *Sagittarius*. Full Moon is on August 15, on the border of *Capricornus* and *Aquarius*. On August 30, at New Moon, the Moon, the Sun, *Mercury*, *Venus* and *Mars* are all clustered together in Leo.

The planets

Mercury is at greatest eastern elongation on August 9, but rapidly becomes close to the Sun. *Venus* is invisible near the Sun in daylight. It is at superior conjunction on August 14. *Mars* (in *Leo*) at mag. 1.8 is also too close to the Sun to be readily visible, but might be glimpsed early in the month as it sets in the west. *Jupiter* (mag. -2.4 to -2.2) is in *Ophiuchus*, reaches a stationary point on August 8 and then begins normal eastwards motion. *Saturn* (mag. 0.1 to 0.3) is still retrograding slowly in *Sagittarius*. *Uranus* (mag. 5.9) and *Neptune* (mag. 7.9) remain in *Aries* and *Aquarius*, respectively.

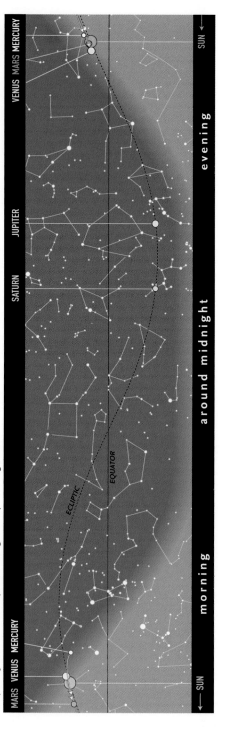

The path of the Sun and the planets along the ecliptic in August.

Calendar for August

01	03:12	New Moon
01	19:55	Mars 1.7°S of Moon
02	07:11	Moon at perigee (359,398 km)
02	11:41	Regulus 3.2°S of Moon
06	00:35	Spica 7.8°S of Moon
07	17:31	First Quarter
09	12:50	Antares 7.9°S of Moon
09	22:53	Jupiter 2.5°S of Moon
09	23:08	Mercury at greatest elongation (19.0°W, mag. -0.0)
11–12		Perseid shower maximum
12	09:53	Saturn 0.1°S of Moon (Occultation from NZ and E. Australia)
14	06:07	Venus at superior conjunction
15	12:29	Full Moon
17	10:49	Moon a0f Regulus
21	04:00 *	Venus 1.0°N of Regulus
23	14:56	Last Quarter
24	09:54	Aldebaran 2.4°S of Moon
27	14:56	Pollux 6.1°N of Moon
28		Alpha Aurigid shower maximum
29	22:19	Regulus 3.2°S of Moon
30	01:07	Mercury 1.9°S of Moon
30	10:22	Mars 3.1°S of Moon
30	10:37	New Moon
30	15:53	Moon at perigee (357,176 km)
30	16:18	Venus 2.9°S of Moon

* These objects are close together for an extended period around this time.

Evening 21:30 (BST)

August 5–6 • The Moon is near Spica, low in the southwest.

Evening 22:30 (BST)

August 8–9 • The Moon with Jupiter, Antares and Sabik.

Evening 22:30 (BST)

August 10–12 • The Moon moves away from Jupiter to pass Nunki and Saturn, low in the southern sky.

Early morning 3:00 (BST)

August 24 • The Moon is between Aldebaran and the Pleiades.

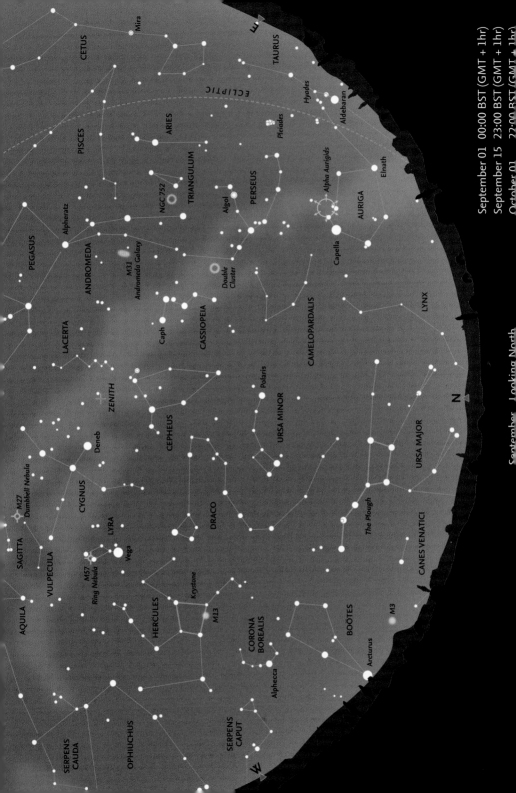

September 01 00:00 BST (GMT + 1hr)
September 15 23:00 BST (GMT + 1hr)
October 01 22:00 BST (GMT + 1hr)

September Looking North

September – Looking North

Ursa Major is now low in the north and to the northwest *Arcturus* and much of *Boötes* sink below the horizon later in the night and later in the month. In the northeast *Auriga* is beginning to climb higher in the sky. Later in the month, *Taurus*, with orange *Aldebaran* (α Tauri), and even *Gemini*, with *Castor* and *Pollux*, become visible in the east and northeast. Due east, *Andromeda* is now clearly visible, with the small constellations of *Triangulum* and *Aries* (the latter a zodiacal constellation) directly below it. Practically the whole of the Milky Way is visible, arching across the sky, both in the north and in the south. It is not particularly clear in Auriga, or even *Perseus*, but in *Cassiopeia* and on towards *Cygnus* the clouds of stars become easier to see. The *Double Cluster* in Perseus is well placed for observation. Cepheus is 'upside-down' near the zenith, and the head of *Draco* and *Hercules* (with the globular cluster *M13*), beyond it are well placed for observation.

Meteors

After the major Perseid shower in August, there is very little shower activity in September. One minor, but very extended, shower, known as the *Alpha Aurigids*, actually has two peaks of activity. The first was on August 28, but the primary peak occurs on September 15. At either of the maxima, however, the hourly rate hardly reaches 10 meteors per hour, although the meteors are bright and relatively easy to photograph. Activity from this shower also extends into October. The *Southern Taurid* shower begins this month and, although rates are low, often produces very bright fireballs. As a slight compensation for the lack of activity, however, in September the number of sporadic meteors reaches its highest rate than at any other time during the year.

The twin open clusters, known as the Double Cluster, in Perseus (more formally called h and χ Persei) are close to the main portion of the Milky Way.

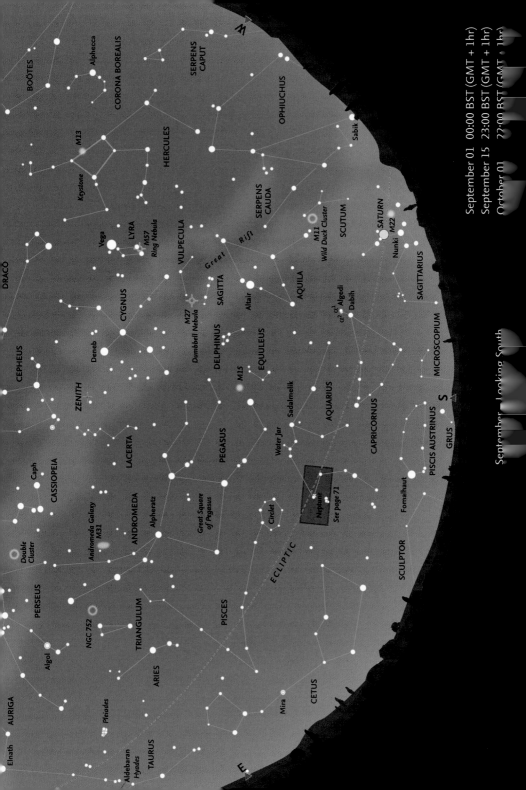

September 01 00:00 BST (GMT + 1hr)
September 15 23:00 BST (GMT + 1hr)
October 01 22:00 BST (GMT + 1hr)

September – Looking South

N

BOÖTES
Alphecca
CORONA BOREALIS
SERPENS CAPUT
OPHIUCHUS
Sabik
M13
HERCULES
SATURN
Keystone
M11
Wild Duck Cluster
SCUTUM
M22
SERPENS CAUDA
LYRA
Vega
M57
Ring Nebula
VULPECULA
Great Rift
Nunki
SAGITTARIUS
DRACO
CYGNUS
SAGITTA
M27
Dumbbell Nebula
AQUILA
Altair
α² Algedi
α¹ Dabih
MICROSCOPIUM
Deneb
DELPHINUS
EQUULEUS
M15
CEPHEUS
ZENITH
AQUARIUS
CAPRICORNUS
S
PISCIS AUSTRINUS
GRUS
LACERTA
PEGASUS
Sadalmelik
Caph
CASSIOPEIA
Circlet
Water Jar
Fomalhaut
ANDROMEDA
Alpheratz
Great Square
of Pegasus
Neptune
See page 71
Double
Cluster
Andromeda Galaxy
M31
SCULPTOR
PERSEUS
NGC 752
TRIANGULUM
PISCES
ECLIPTIC
Algol
ARIES
CETUS
Mira
AURIGA
Elnath
Pleiades
Aldebaran
Hyades
TAURUS

E

September – Looking South

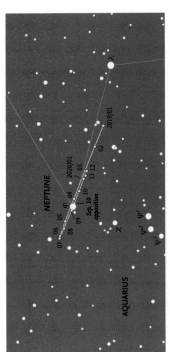

The **Summer Triangle** is now high in the southwest, with the Great Square of **Pegasus** high in the southeast. Below Pegasus are the two zodiacal constellations of **Capricornus** and **Aquarius**. In what is otherwise an unremarkable constellation, **Algedi** (α Capricorni) is actually a visual binary, with the two stars (α¹ Cap and α² Cap) readily seen with the naked eye. **Dabih** (β Capricorni), just to the south, is also a double star, and the components are relatively easy to separate with binoculars. In Aquarius, just to the east of **Sadalmelik** (α Aquarii) there is a small asterism consisting of four stars, resembling a tiny letter 'Y', known as the 'Water Jar'. Below Aquarius is a sparsely populated area of the sky with just one bright star in the constellation of **Piscis Austrinus.** In classical illustrations, water is shown flowing from the 'Water Jar' towards bright **Fomalhaut** (α Piscis Austrini).

Another zodiacal constellation, **Pisces**, is now clearly visible to the east of Aquarius. Although faint, there is a distinctive asterism of stars, known as the 'Circlet', south of the Great Square and another line of faint stars to the east of Pegasus. Still farther down towards the horizon

The path of Neptune in 2019. Neptune comes to opposition on September 10. All stars brighter than magnitude 8.5 are shown.

is the constellation of **Cetus**, with the famous variable star **Mira** (ο Ceti) at its centre. When Mira is at maximum brightness (around mag. 3.5) it is clearly visible to the naked eye, but it disappears as it fades towards minimum (about mag. 9.5 or less). There is a finder chart for Mira on page 83.

The Moon's phases for September

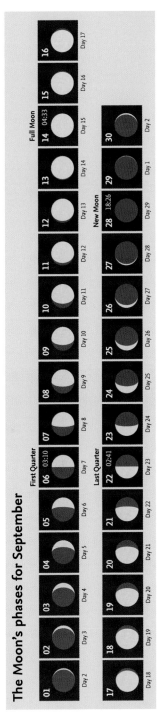

September – Moon and Planets

The Moon

On September 5, one day before First Quarter, the Moon is close to **Antares** and, the next day, near **Jupiter** in **Ophiuchus**. On September 8, 9 days old, it passes **Saturn** in **Sagittarius**. (An occultation is visible from Australia.) Full Moon (in **Pisces**) is on September 14. By September 20, as waning gibbous, the Moon is close to **Aldebaran** in **Taurus**. On September 26, the Moon is close to **Regulus** in **Leo**, visible in the early morning before sunrise. New Moon is on September 28 when the Moon (and the Sun) are in **Virgo**.

The planets

Mercury and **Venus** are too close to the Sun and too low to be visible. **Mars** is also lost in daylight. **Jupiter** (mag. -2.2 to -2.0) is moving eastwards slowly in **Ophiuchus**. **Saturn** (mag. 0.3 to 0.5) is also moving eastwards slowly in **Sagittarius**. **Uranus** (mag. 5.9) is in **Aries**. **Neptune** (in **Aquarius**) comes to opposition on September 10 at mag. 7.8 (on page 71 there is a finder chart showing the path of Neptune in 2019).

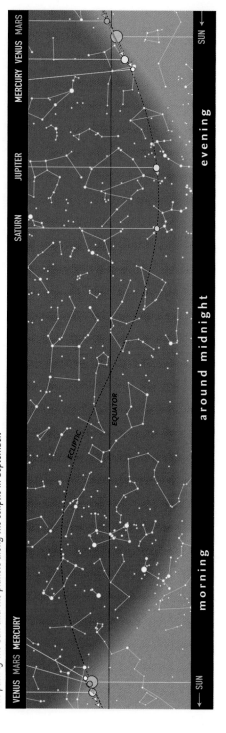

The path of the Sun and the planets along the ecliptic in September.

Calendar for September

02	09:13	Spica 7.7°S of Moon
02	10:42	Mars in conjunction with Sun
04	01:40	Mercury at superior conjunction
05	19:06	Antares 7.8°S of Moon
06	03:10	First Quarter
06	06:52	Jupiter 2.3°S of Moon
08	13:42	Saturn 0.1°N of Moon
10	07:24	Neptune at opposition (mag. 7.8)
13	13:32	Moon at apogee (406,377 km)
14	04:33	Full Moon
15		Alpha Aurigid shower second maximum
20	16:45	Aldebaran 2.7°S of Moon
22	02:41	Last Quarter
23	07:50	Autumnal equinox
23–Nov.19		Southern Taurid shower
23–Nov.27		Orionid meteor shower
24	00:01	Pollux 3.3°N of Moon
26	08:54	Regulus 3.3°S of Moon
28	01:19	Mars 4.1°S of Moon
28	02:24	Moon at perigee (357,802 km)
28	18:26	New Moon
29	12:46	Mars 4.4°S of Moon
29	19:44	Spica 7.6°S of Moon
29	22:01	Mercury 6.2°S of Moon

Evening 21:00 (BST)

September 5–8 • *The Moon passes Antares, Jupiter, Sabik, Nunki and Saturn.*

Evening 23:30 (BST)

September 19–20 • *The Moon passes Aldebaran, 10 degrees below the Pleiades.*

After midnight 2:00 (BST)

September 23–24 • *The Moon passes Alhena and one day later it almost lines up with Castor and Pollux.*

Morning 6:00 (BST)

September 26 • *The Moon with Regulus and Algieba. Denebola is close to the horizon.*

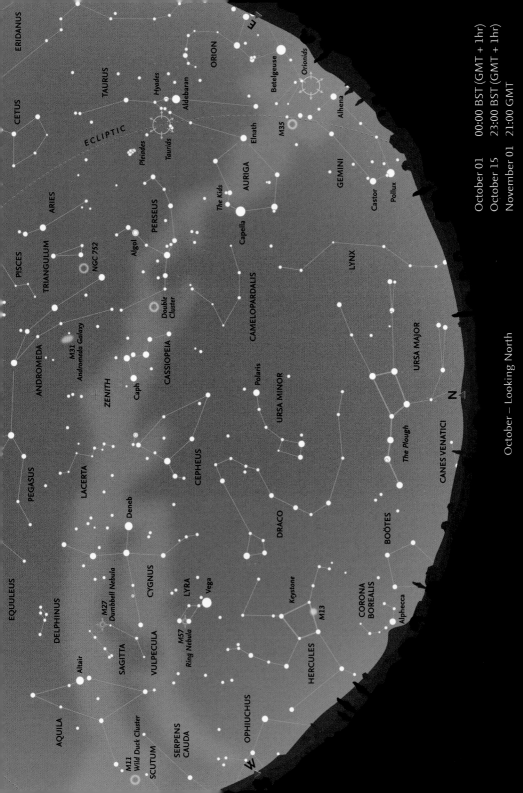

October 01 00:00 BST (GMT + 1hr)
October 15 23:00 BST (GMT + 1hr)
November 01 21:00 GMT

October – Looking North

October – Looking North

Ursa Major is grazing the horizon in the north, while high overhead are the constellations of *Cepheus*, *Cassiopeia* and *Perseus*, with the Milky Way between Cepheus and Cassiopeia at the zenith. *Auriga* is now clearly visible in the east, as is *Taurus* with the *Pleiades*, *Hyades* and orange *Aldebaran*. Also in the east, *Orion* and *Gemini* are starting to rise clear of the horizon.

The constellations of *Boötes* and *Corona Borealis* are now essentially lost to view in the northwest, and *Hercules* is also descending towards the western horizon. The three stars of the Summer Triangle are still clearly visible, although *Aquila* and *Altair* are beginning to approach the horizon in the west. Towards the end of the month (October 27) Summer Time ends in Europe, with Britain reverting to Greenwich Mean Time and Europe to Central European Time.

Meteors

The *Orionids* are the major, fairly reliable meteor shower active in October. Like the May *Eta Aquariid* shower, the Orionids are associated with Comet 1P/Halley. During this second pass through the stream of particles from the comet, slightly fewer meteors are seen than in May, but conditions are more favourable for northern observers. In both showers the meteors are very fast, and many leave persistent trains. Although the Orionid maximum is quoted as October 21–22, in fact there is a very broad maximum, lasting about a week from October 20 to 27, with hourly rates around 25. Occasionally rates are higher (50–70 per hour). In 2019, the broad maximum extends from the day before Last Quarter to a waning crescent, so moonlight should not cause too much interference.

The faint shower of the *Southern Taurids* (often with bright fireballs) peaks on October 28–29. The Southern Taurid maximum occurs when the Moon is a waning crescent, so conditions are generally favourable. Towards the end of the month (around October 19), another shower (the *Northern Taurids*) begins to show activity, which peaks early in November. The parent comet for both Taurid showers is Comet 2P/Encke.

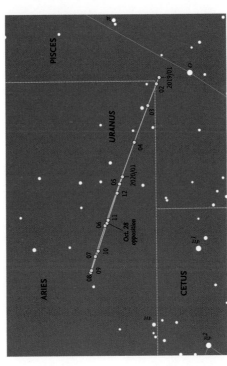

The path of Uranus in 2019. Uranus comes to opposition on October 28. All stars brighter than magnitude 7.5 are shown.

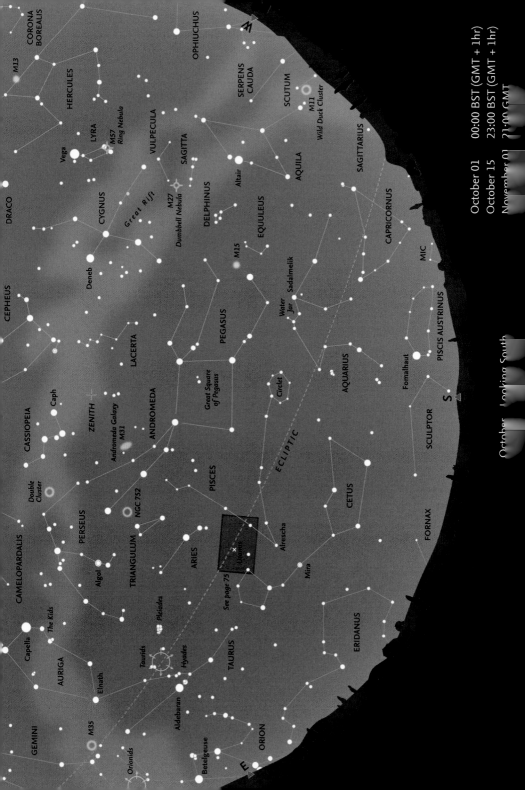

CORONA
BOREALIS

M13

OPHIUCHUS

DRACO

HERCULES

SERPENS
CAUDA

M57
Ring Nebula

LYRA

SCUTUM

Vega

M11
Wild Duck Cluster

CEPHEUS

CYGNUS

VULPECULA

SAGITTA

SAGITTARIUS

Deneb

M27
Dumbbell Nebula

DELPHINUS

AQUILA

Great Rift

Altair

EQUULEUS

CAPRICORNUS

M15

LACERTA

PEGASUS

Water
Jar

Sadalmelik

MIC

Caph

ZENITH

CASSIOPEIA

Great Square
of Pegasus

AQUARIUS

PISCIS AUSTRINUS

Andromeda Galaxy
M31

ANDROMEDA

Circlet

Fomalhaut

Double
Cluster

PISCES

ECLIPTIC

SCULPTOR

S

PERSEUS

NGC 752

CETUS

Algol

TRIANGULUM

See page 75

Uranus

CAMELOPARDALIS

ARIES

Alrescha

FORNAX

The Kids

Pleiades

Mira

Capella

Taurids

Hyades

ERIDANUS

AURIGA

Elnath

TAURUS

Aldebaran

M35

GEMINI

Orionids

Betelgeuse

ORION

E

October 01   00:00 BST (GMT + 1hr)
October 15   23:00 BST (GMT + 1hr)
November 01   21:00 GMT

October — Looking South

October – Looking South

The Great Square of **Pegasus** dominates the southern sky, framed by the two chains of stars that form the constellation of **Pisces**, together with **Alrescha** (α Piscium) at the point where the two lines of stars join. Also clearly visible is the constellation of **Cetus**, below Pegasus and Pisces. Although **Capricornus** is beginning to disappear, **Aquarius** to its east is well placed in the south, with solitary **Fomalhaut** and the constellation of **Piscis Austrinus** beneath it, close to the horizon.

The main band of the Milky Way and the Great Rift runs down from **Cygnus**, through **Vulpecula**, **Sagitta** and **Aquila** towards the western horizon. **Delphinus** and the tiny, unremarkable constellation of **Equuleus** lie between the band of the Milky Way and Pegasus. **Andromeda** is clearly visible high in the sky to the southeast, with the small constellation of **Triangulum** and the zodiacal constellation of **Aries** below it. **Perseus** is high in the east, and by now the **Pleiades** and **Taurus** are well clear of the horizon. Later in the night, and later in the month, **Orion** rises in the east, a sign that the autumn season has arrived and of the steady approach of winter.

The Moon's phases for October

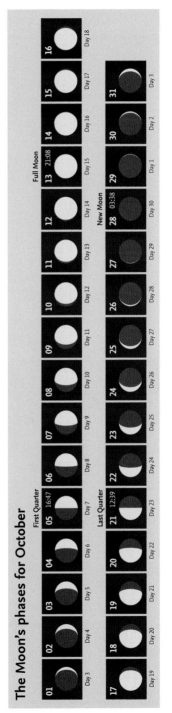

The constellation of Aquarius is one of the constellations that is visible in late summer and early autumn. The four stars forming the 'Y'-shape of the 'Water Jar' may be seen to the east of Sadalmelik (α Aquarii), the brightest star (top centre).

October – Moon and Planets

The Moon

The Moon is near **Antares** on October 3, but is visible only early just before setting. Later that day it is close to **Jupiter** and both bodies may be glimpsed just before they set in the west. On October 5, the Moon passes very close to **Saturn**. (There is an occultation visible from southern Africa.) Full Moon is on October 13. On October 17, the waning gibbous Moon is close to **Aldebaran** in **Taurus** and on October 23 it is near **Regulus** in **Leo**, but the constellation rises only later the next morning. New Moon occurs on October 28.

The planets

Mercury comes to greatest eastern elongation on October 20, but is invisible in daylight in **Libra**. **Venus** remains close to the Sun, also in Libra and not visible. **Mars** (in **Virgo**) is similarly too close to the Sun to be seen. **Jupiter** (mag. -2.0 to -1.9) is moving slowly eastwards in **Ophiuchus** and is visible in the early evening before it sets, as is **Saturn** in **Sagittarius** at mag. 0.5. **Uranus** remains in **Aries**, and comes to opposition on October 28 at mag. 5.7 (on page 75 there is a finder chart showing the path of Uranus in 2019). **Neptune** is in **Aquarius** at mag. 7.9. Both constellations are visible for a large part of the night.

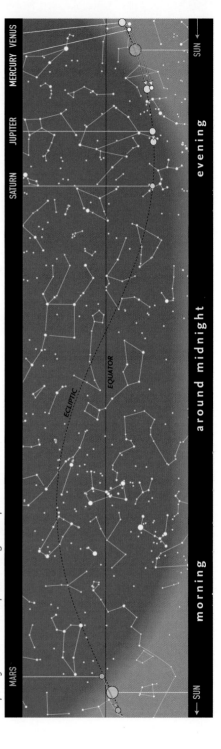

The path of the Sun and the planets along the ecliptic in October.

Calendar for October

03	01:00 *	Venus 3.1°N of Spica
03	03:18	Antares 7.5°S of Moon
03	20:23	Jupiter 1.9°S of Moon
05	16:47	First Quarter
05	20:36	Saturn 0.3°N of Moon (Occultation from S. Africa)
10	18:29	Moon at apogee (405,898 km)
13	21:08	Full Moon
17	22:22	Aldebaran 2.9°S of Moon
19–Dec.10		Northern Taurid meteor shower
20	04:02	Mercury at greatest elongation (24.6°E, mag. -0.1)
21	06:49	Pollux 5.6°N of Moon
21	12:39	Last Quarter
21-22		Orionid shower maximum
23	17:37	Regulus 3.5°S of Moon
26	10:39	Moon at perigee (363,101 km)
26	16:52	Mars 4.5°S of Moon
27	06:30	Spica 7.6°S of Moon
27		Summer Time ends (Europe)
28	03:38	New Moon
28	08:15	Uranus at opposition (mag. 5.7)
28–29		Southern Taurid shower maximum
29	14:55	Mercury 6.7°S of Moon
29	13:32	Venus 3.9°S of Moon
30	13:14	Antares 7.3°S of Moon
31	14:22	Jupiter 1.3°S of Moon

* These objects are close together for an extended period around this time.

Evening 22:30 (BST)

October 17 • The Moon and Aldebaran with the Pleiades twelve degrees higher.

Evening 19:30 (BST)

October 2–5 • The Moon passes Sabik, Jupiter, Nunki and Saturn. Antares is becoming too close to the horizon to be readily seen.

Morning 7:15 (BST)

October 26–27 • The Moon with Mars, Porrima (γ Vir) and Vindemiatrix (ε Vir), shortly before sunrise.

Morning 5:00 (BST)

October 23–24 • The Moon passes Regulus and Algieba.

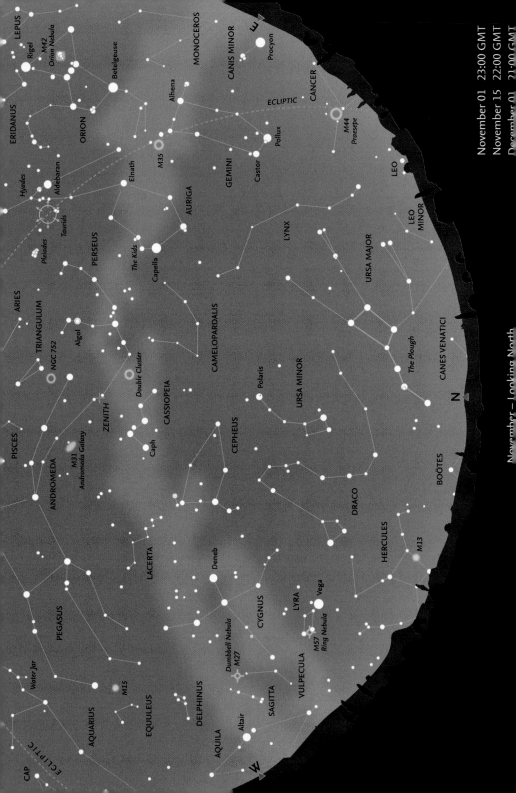

November 01   23:00 GMT
November 15   22:00 GMT
December 01   21:00 GMT

November – Looking North

November – Looking North

Most of **Aquila** has now disappeared below the horizon, but two of the stars of the Summer Triangle, **Vega** in **Lyra** and **Deneb** in **Cygnus**, are still clearly visible in the west. The head of **Draco** is now low in the northwest and only a small portion of **Hercules** remains above the horizon. The southernmost stars of **Ursa Major** are now coming into view. The Milky Way arches overhead, with the denser star clouds in the west and the less heavily populated region through **Auriga** and **Monoceros** in the east. High overhead, **Cassiopeia** is near the zenith and **Cepheus** has swung round to the north, while Auriga is now high in the northeast. **Gemini**, with **Castor** and **Pollux**, is well clear of the eastern horizon, and even **Procyon** (α Canis Minoris) is just climbing into view almost due east.

At the beginning of the month (November 3) Daylight Saving Time comes to an end in North America.

Meteors

The **Northern Taurid** shower, which began in mid-October, reaches maximum – although with only a low rate of about five meteors per hour – on November 12, so conditions are extremely unfavourable. The shower gradually trails off, ending around December 10. There is an apparent 7-year periodicity in fireball activity, but 2019 is unlikely to be another peak year. Far more striking, however, are the **Leonids**, which have a relatively short period of activity (November 5–30), with maximum on November 17–18. This shower is associated with Comet 55P/Tempel-Tuttle and has shown extraordinary activity on various occasions with many thousands of meteors per hour. High rates were seen in 1999, 2001 and 2002 (reaching about 3000 meteors per hour) but have fallen dramatically since then. The rate in 2019 is likely to be about 15 per hour. These meteors are the fastest shower meteors recorded (about 70 km per second) and often leave persistent trains. The shower is very rich in faint meteors. In 2019, maximum is about a day before Last Quarter, when the Moon is waning gibbous, so conditions are not particularly favourable.

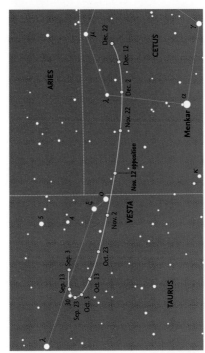

A finder chart for minor planet Vesta (4) which is at opposition (mag. 6.5) on November 12. Background stars are shown down to magnitude 7.5.

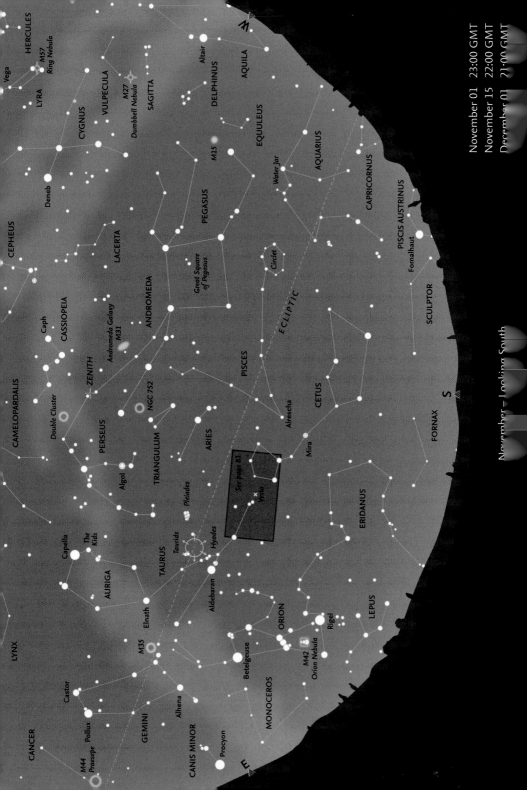

November 01 23:00 GMT
November 15 22:00 GMT
December 01 21:00 GMT

November – Looking South

November – Looking South

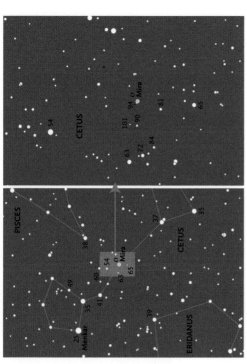

Orion has now risen above the eastern horizon, and part of the long, straggling constellation of *Eridanus* (which begins near *Rigel*) is visible to the west of Orion. Higher in the sky, *Taurus*, with the *Pleiades* cluster, and orange *Aldebaran* are now easy to observe. To their west, both *Pisces* and *Cetus* are close to the meridian. The famous long-period variable star, *Mira* (o Ceti), with a typical range of magnitude 3.4 to 9.8, is favourably placed for observation. In the southwest, *Capricornus* has slipped below the horizon, but *Aquarius* remains visible. Even farther west, *Altair* may be seen early in the night, but most of *Aquila* has already disappeared from view. *Delphinus*, together with *Sagitta* and *Vulpecula* in the Milky Way, will soon vanish for another year. Both *Pegasus* and *Andromeda* are easy to see, and one of the lines of stars that make up Andromeda finishes close to the zenith, which is also close to one of the outlying stars of *Perseus*, high in the east.

Finder and comparison charts for Mira (o Ceti). The chart on the left shows all stars brighter than magnitude 6.5. The chart on the right shows stars down to magnitude 10.0. The comparison star magnitudes are shown without the decimal point.

The Moon's phases for November

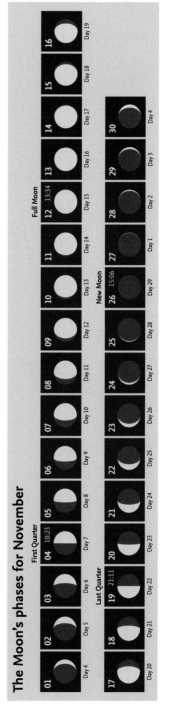

First Quarter

| 01 | 02 | 03 | 04 10:23 | 05 | 06 | 07 | 08 | 09 | 10 | 11 | 12 13:34 |
| Day 4 | Day 5 | Day 6 | Day 7 | Day 8 | Day 9 | Day 10 | Day 11 | Day 12 | Day 13 | Day 14 | Day 15 |

Full Moon

| 13 | 14 | 15 | 16 |
| Day 16 | Day 17 | Day 18 | Day 19 |

Last Quarter

| 17 | 18 | 19 21:11 | 20 | 21 | 22 | 23 | 24 | 25 | 26 15:06 | 27 | 28 | 29 | 30 |
| Day 20 | Day 21 | Day 22 | Day 23 | Day 24 | Day 25 | Day 26 | Day 27 | Day 28 | Day 29 | Day 1 | Day 2 | Day 3 | Day 4 |

New Moon

November – Moon and Planets

The Moon

On November 2 the Moon (a waxing crescent, two days before First Quarter) is very close to **Saturn** in **Sagittarius**. (There is an occultation, visible from New Zealand and the southern tip of Tasmania.) Full Moon is on November 12, when the Moon is on the border of **Aries** with **Taurus**. Two days later, on November 14, the Moon is north of **Aldebaran**. On November 19, at Last Quarter, it is near **Regulus** in **Leo**. On November 23, a waning crescent, it passes **Spica** in **Virgo** and, a day later, is close to **Mars**. The three bodies are visible in the early-morning sky. New Moon is on November 26, when it is in the small section of **Scorpius** between **Libra** and **Ophiuchus**. Two days later it is close to **Jupiter** in the morning sky, and on November 29 it is again close to **Saturn**.

The planets

Mercury is initially close to the Sun, passing inferior conjunction on November 11, but rapidly moves to greatest western elongation on November 28, when it may be glimpsed in the morning sky. **Venus** (mag. -3.8 to -3.9) is very low in the evening sky. **Mars** (mag. 1.8–1.7) is in **Libra**, visible in the early morning. **Jupiter** (mag. -1.9 to -1.8), initially in **Ophiuchus**, moves into **Sagittarius** and is visible in the southwest just before it sets. **Saturn** (mag. 0.5–0.6) is also in **Sagittarius**. **Uranus** is mag. 5.7 in **Aries** and **Neptune** (mag. 7.9) is in **Aquarius**. The minor planet (4) **Vesta** comes to opposition in **Cetus** on November 12. There is a finder chart on page 81.

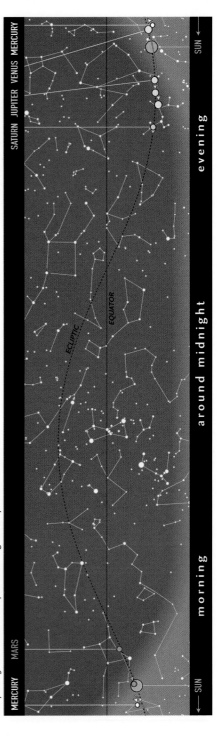

The path of the Sun and the planets along the ecliptic in November.

Calendar for November

02	07:21	Saturn 0.6°N of Moon
03		Daylight Saving Time ends in North America
04	10:23	First Quarter
05–30		Leonid meteor shower
07	08:36	Moon at apogee (405,058 km)
08	15:00 *	Mars 3.0°N of Spica
09	11:00 *	Venus 4.0°N of Antares
11	15:22	Mercury at inferior conjunction
12	08:56	Vesta at opposition (mag. 6.5)
12	13:34	Full Moon
14	04:23	Aldebaran 3.0°S of Moon
17	12:10	Pollux 5.4°N of Moon
17–18		Leonid shower maximum
19	21:11	Last Quarter
19	23:51	Regulus 3.7°S of Moon
23	07:41	Moon at perigee (366,716 km)
23	15:32	Spica 7.7°S of Moon
24	09:02	Mars 4.3°S of Moon
24	14:00 *	Venus 1.4°S of Jupiter
25	02:50	Mercury 1.9°S of Moon
26	15:06	New Moon
26	23:29	Antares 7.2°S of Moon
28	10:29	Mercury at greatest elongation (20.1°W, mag. −0.6)
28	10:49	Jupiter 0.7°S of Moon
28	18:49	Venus 1.9°S of Moon
29	21:03	Saturn 0.9°N of Moon

* These objects are close together for an extended period around this time.

Evening 17:30

October 30 – November 2 • *The Moon with Sabik, Jupiter, Nunki and Saturn, low in the southwest, shortly after sunset.*

Morning 6:30

November 9 • *Mars and Spica in the morning sky.*

Morning 6:30

November 23–25 • *The Moon passes Spica, Mars and Mercury in the southeast.*

Evening 16:45

November 28–29 • *The Moon in the company of Jupiter, Venus and Saturn, very low in the southwest.*

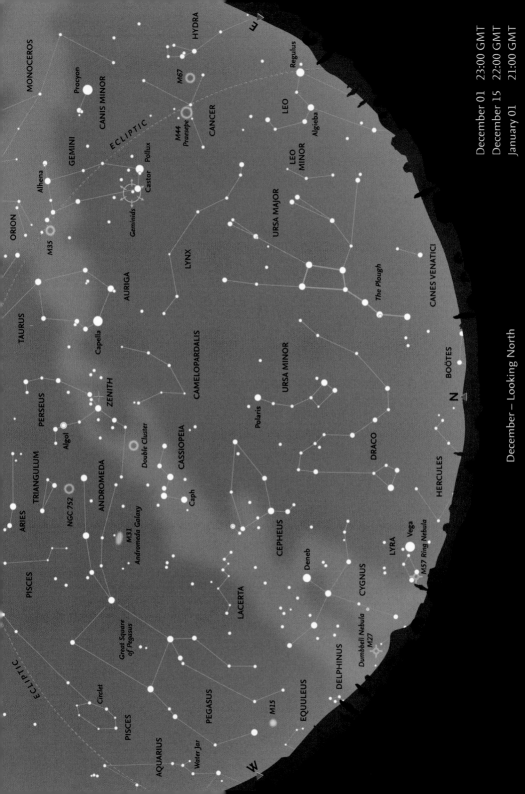

December 01   23:00 GMT
December 15   22:00 GMT
January 01   21:00 GMT

December – Looking North

December – Looking North

Ursa Major has now swung around and is starting to 'climb' in the east. The fainter stars in the southern part of the constellation are now fully in view. The other bear, *Ursa Minor*, 'hangs' below *Polaris* in the north. Directly above it is the faint constellation of *Camelopardalis*, with the other inconspicuous circumpolar constellation, *Lynx*, to its east. *Vega* (α Lyrae) is skimming the horizon in the northwest, but *Deneb* (α Cygni) and most of *Cygnus* remain visible farther west. In the east, *Regulus* (α Leonis) and the constellation of *Leo* are beginning to rise above the horizon. *Cancer* stands high in the east, with *Gemini* even higher in the sky. *Perseus* is at the zenith, with *Auriga* and *Capella* between it and Gemini. Because it is so high in the sky, now is a good time to examine the star clouds of the fainter portion of the Milky Way, between *Cassiopeia* in the west to Gemini and *Orion* in the east.

Meteors

There is one significant meteor shower in December (the last major shower of the year). This is the *Geminid* shower, which is visible over the period December 4–16 and comes to maximum on December 13–14, just after Full Moon (December 12) so conditions are very poor. It is one of the most active showers of the year, and in some years is the strongest, with a peak rate of around 100 meteors per hour. It is the one major shower that shows good activity before midnight. The meteors have been found to have a much higher density than other meteors (which are derived from cometary material). It was eventually established that the Geminids and the asteroid Phaeton had similar orbits. So the Geminids are assumed to consist of denser, rocky material. They are slower than most other meteors and often appear to last longer. The brightest often break up into numerous luminous fragments that follow similar paths across the sky. There is a second shower: the *Ursids*, active December 17–23, peaking on December 21–22, with rate at maximum of 5–10, occasionally rising to 25 per hour. Maximum in 2019 is when the Moon is a waning crescent, so conditions are reasonably favourable. The parent body is Comet 8P/Tuttle.

The constellation of Cassiopeia is a familiar sight among the northern circumpolar constellations. It is always above the horizon, on the opposite side of Polaris (the North Star) to the equally well-known asterism of the seven stars of the Plough.

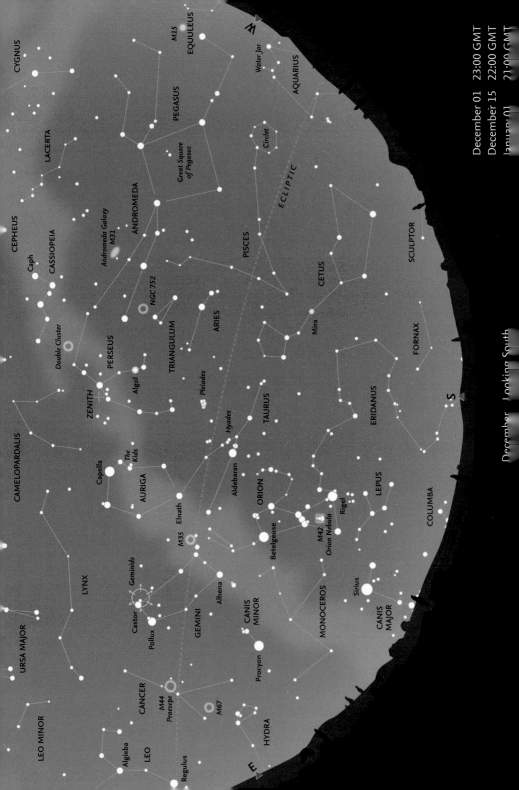

CYGNUS

M15

EQUULEUS

Water Jar

PEGASUS

AQUARIUS

CEPHEUS

LACERTA

Great Square
of Pegasus

Circlet

ECLIPTIC

ANDROMEDA

Caph

CASSIOPEIA

Andromeda Galaxy
M31

PISCES

NGC 752

SCULPTOR

CETUS

Double Cluster

PERSEUS

TRIANGULUM

ARIES

Mira

FORNAX

ZENITH

Algol

Pleiades

TAURUS

ERIDANUS

CAMELOPARDALIS

Hyades

Capella

The
Kids

Aldebaran

S

Elnath

AURIGA

ORION

LEPUS

LYNX

M35

Betelgeuse

Rigel

COLUMBA

M42
Orion Nebula

Geminids

Castor

Alhena

MONOCEROS

Sirius

URSA MAJOR

Pollux

GEMINI

CANIS
MINOR

CANIS
MAJOR

Procyon

CANCER

M44
Praesepe

LEO MINOR

M67

Algieba

LEO

HYDRA

E

Regulus

December 01   23:00 GMT
December 15   22:00 GMT
January 01   21:00 GMT

December Looking South

December – Looking South

The fine open cluster of the *Pleiades* is due south around 22:00, with the *Hyades* cluster, *Aldebaran* and the rest of *Taurus* clearly visible to the east. *Auriga* (with *Capella*) and *Gemini* (with *Castor* and *Pollux*) are both well-placed for observation. *Orion* has made a welcome return to the winter sky, and both *Canis Minor* (with *Procyon*) and *Canis Major* (with *Sirius*, the brightest star in the sky) are now well above the horizon. The small, poorly known constellation of *Lepus* lies to the south of Orion. In the west, *Aquarius* has now disappeared, and *Cetus* is becoming lower, but *Pisces* is still easily seen, as are the constellations of *Aries*, *Triangulum* and *Andromeda* above it. The Great Square of *Pegasus* is starting to plunge down towards the western horizon, and because of its orientation on the sky appears more like a large diamond, standing on one point, than a square.

The constellation of Taurus contains two contrasting open clusters: the compact Pleiades, with its striking blue-white stars, and the more scattered, 'V'-shaped Hyades, which are much closer to us. Orange Aldebaran (α Tauri) is not related to the Hyades, but lies between it and the Earth.

The Moon's phases for December

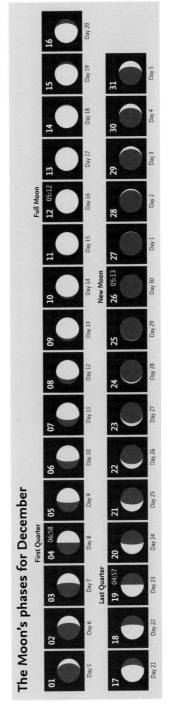

First Quarter

01	02	03	04 06:58	05	06	07	08	09
Day 5	Day 6	Day 7	Day 8	Day 9	Day 10	Day 11	Day 12	Day 13

Last Quarter   Full Moon

10	11	12 05:12	13	14	15	16		
Day 14	Day 15	Day 16	Day 17	Day 18	Day 19	Day 20		

New Moon

17	18	19 04:57	20	21	22	23	24	25
Day 21	Day 22	Day 23	Day 24	Day 25	Day 26	Day 27	Day 28	Day 29

26 05:13	27	28	29	30	31	
Day 30	Day 1	Day 2	Day 3	Day 4	Day 5	

December – Moon and Planets

The Moon

The Moon is close to **Aldebaran** in **Taurus** on December 11, but Full Moon occurs the next day, so the star is overpowered by moonlight. On December 17, as waning gibbous, two days before Last Quarter, it is near **Regulus** in **Leo**, and on December 20, one day after Last Quarter, it is fairly close to **Spica** in **Virgo**. On December 23, as a waning crescent, it is close to **Mars** in **Libra**. At New Moon on December 26 there is an annular eclipse of the Sun, visible from Arabia, the Indian Ocean and Indonesia.

The planets

Mercury is rapidly approaching the Sun and is not visible. **Venus** begins the month low in the southwestern evening sky at mag. -3.9, and rapidly moves east into **Capricornus**, becoming more easily seen and brightening slightly. **Mars** (mag. 1.7) begins the month just inside **Virgo** and moves east into **Libra**. **Jupiter** is close to the Sun in **Sagittarius**, and **Saturn** is also in that constellation. **Uranus** is still just inside **Aries** near the border with **Pisces**, and **Neptune** remains in **Aquarius** where it has been throughout the year.

The path of the Sun and the planets along the ecliptic in December.

Calendar for December

04	06:58	First Quarter
04–16		Geminid meteor shower
05	04:08	Moon at apogee (404,446 km)
11	12:10	Aldebaran 3.0°S of Moon
12	05:12	Full Moon
13–14		Geminid shower maximum
14	18:19	Pollux 5.3°N of Moon
17–23		Ursid meteor shower
17	05:08	Regulus 3.8°S of Moon
18	20:25	Moon at perigee (370,265 km)
19	04:57	Last Quarter
20	22:03	Spica 7.8°S of Moon
21–22		Ursid shower maximum
22	04:19	Winter solstice
23	01:49	Mars 3.5°S of Moon
24	08:12	Antares 7.7°S of Moon
25	11:08	Mercury 1.9°S of Moon
26	05:13	New Moon
26	07:30	Jupiter 0.2°S of Moon
26	05:17	Annular solar eclipse
		(Arabia, Indian Ocean, Indonesia)
27	11:47	Saturn 1.2°N of Moon
27	18:26	Jupiter in conjunction with Sun
29	01:31	Venus 1.0°N of Moon

Evening 18:00

December 10–11 • *The Moon passes Aldebaran and the Pleiades, almost due east.*

Morning 6:30

December 17 • *The Moon lines up with Regulus and Algieba, high in the southwest.*

Morning 7:30

December 22–24 • *The Moon passes Mars and Antares, just before sunrise. Antares is probably too low to be readily seen.*

Evening 17:30

December 29 • *The Moon with Venus in the southwest, after sunset.*

Glossary and Tables

aphelion	The point on an orbit that is farthest from the Sun.
apogee	The point on its orbit at which the Moon is farthest from the Earth.
appulse	The apparently close approach of two celestial objects; two planets, or a planet and star.
astronomical unit	(AU) The mean distance of the Earth from the Sun, 149,597,870 km.
celestial equator	The great circle on the celestial sphere that is in the same plane as the Earth's equator.
celestial sphere	The apparent sphere surrounding the Earth on which all celestial bodies (stars, planets, etc.) seem to be located.
conjunction	The point in time when two celestial objects have the same celestial longitude. In the case of the Sun and a planet, superior conjunction occurs when the planet lies on the far side of the Sun (as seen from Earth). For Mercury and Venus, inferior conjuction occurs when they pass between the Sun and the Earth.
direct motion	Motion from west to east on the sky.
ecliptic	The apparent path of the Sun across the sky throughout the year. Also: the plane of the Earth's orbit in space.
elongation	The point at which an inferior planet has the greatest angular distance from the Sun, as seen from Earth.
equinox	The two points during the year when night and day have equal duration. Also: the points on the sky at which the ecliptic intersects the celestial equator. The vernal (spring) equinox is of particular importance in astronomy.
gibbous	The stage in the sequence of phases at which the illumination of a body lies between half and full. In the case of the Moon, the term is applied to phases between First Quarter and Full, and between Full and Last Quarter.
inferior planet	Either of the planets Mercury or Venus, which have orbits inside that of the Earth.
magnitude	The brightness of a star, planet or other celestial body. It is a logarithmic scale, where larger numbers indicate fainter brightness. A difference of 5 in magnitude indicates a difference of 100 in actual brightness, thus a first-magnitude star is 100 times as bright as one of sixth magnitude.
meridian	The great circle passing through the North and South Poles of a body and the observer's position; or the corresponding great circle on the celestial sphere that passes through the North and South Celestial Poles and also through the observer's zenith.
nadir	The point on the celestial sphere directly beneath the observer's feet, opposite the zenith.
occultation	The disappearance of one celestial body behind another, such as when stars or planets are hidden behind the Moon.
opposition	The point on a superior planet's orbit at which it is directly opposite the Sun in the sky.
perigee	The point on its orbit at which the Moon is closest to the Earth.
perihelion	The point on an orbit that is closest to the Sun.
retrograde motion	Motion from east to west on the sky.
superior planet	A planet that has an orbit outside that of the Earth.
vernal equinox	The point at which the Sun, in its apparent motion along the ecliptic, crosses the celestial equator from south to north. Also known as the First Point of Aries.
zenith	The point directly above the observer's head.
zodiac	A band, stretching 8° on either side of the ecliptic, within which the Moon and planets appear to move. It consists of twelve equal areas, originally named after the constellation that once lay within it.

The Greek Alphabet

α	Alpha	ε	Epsilon	ι	Iota	ν	Nu	ρ	Rho	φ (φ)	Phi
β	Beta	ζ	Zeta	κ	Kappa	ξ	Xi	σ (ς)	Sigma	χ	Chi
γ	Gamma	η	Eta	λ	Lambda	ο	Omicron	τ	Tau	ψ	Psi
δ	Delta	θ (ϑ)	Theta	μ	Mu	π	Pi	υ	Upsilon	ω	Omega

The Constellations

There are 88 constellations covering the whole of the celestial sphere, but 24 of these in the southern hemisphere can never be seen (even in part) from the latitude of Britain and Ireland, so are omitted from this table. The names themselves are expressed in Latin, and the names of stars are frequently given by Greek letters followed by the genitive of the constellation name. The genitives and English names of the various constellations are included.

Name	Genitive	Abbr.	English name
Andromeda	Andromeda	And	Andromeda
Antlia	Antliae	Ant	Air Pump
Aquarius	Aquarii	Aqr	Water Bearer
Aquila	Aquilae	Aql	Eagle
Aries	Arietis	Ari	Ram
Auriga	Aurigae	Aur	Charioteer
Boötes	Boötis	Boo	Herdsman
Camelopardalis	Camelopardalis	Cam	Giraffe
Cancer	Cancri	Cnc	Crab
Canes Venatici	Canum Venaticorum	CVn	Hunting Dogs
Canis Major	Canis Majoris	CMa	Big Dog
Canis Minor	Canis Minoris	CMi	Little Dog
Capricornus	Capricorni	Cap	Sea Goat
Cassiopeia	Cassiopeiae	Cas	Cassiopeia
Centaurus	Centauri	Cen	Centaur
Cepheus	Cephei	Cep	Cepheus
Cetus	Ceti	Cet	Whale
Columba	Columbae	Col	Dove
Coma Berenices	Comae Berenices	Com	Berenice's Hair
Corona Australis	Coronae Australis	CrA	Southern Crown
Corona Borealis	Coronae Borealis	CrB	Northern Crown
Corvus	Corvi	Crv	Crow
Crater	Crateris	Crt	Cup
Cygnus	Cygni	Cyg	Swan
Delphinus	Delphini	Del	Dolphin
Draco	Draconis	Dra	Dragon
Equuleus	Equulei	Equ	Little Horse
Eridanus	Eridani	Eri	River Eridanus
Fornax	Fornacis	For	Furnace
Gemini	Geminorum	Gem	Twins
Hercules	Herculis	Her	Hercules
Hydra	Hydrae	Hya	Water Snake

Name	Genitive	Abbr.	English name
Lacerta	Lacertae	Lac	Lizard
Leo	Leonis	Leo	Lion
Leo Minor	Leonis Minoris	LMi	Little Lion
Lepus	Leporis	Lep	Hare
Libra	Librae	Lib	Scales
Lupus	Lupi	Lup	Wolf
Lynx	Lyncis	Lyn	Lynx
Lyra	Lyrae	Lyr	Lyre
Microscopium	Microscopii	Mic	Microscope
Monoceros	Monocerotis	Mon	Unicorn
Ophiuchus	Ophiuchi	Oph	Serpent Bearer
Orion	Orionis	Ori	Orion
Pegasus	Pegasi	Peg	Pegasus
Perseus	Persei	Per	Perseus
Pisces	Piscium	Psc	Fishes
Piscis Austrinus	Piscis Austrini	PsA	Southern Fish
Puppis	Puppis	Pup	Stern
Pyxis	Pyxidis	Pyx	Compass
Sagitta	Sagittae	Sge	Arrow
Sagittarius	Sagittarii	Sgr	Archer
Scorpius	Scorpii	Sco	Scorpion
Sculptor	Sculptoris	Scl	Sculptor
Scutum	Scuti	Sct	Shield
Serpens	Serpentis	Ser	Serpent
Sextans	Sextantis	Sex	Sextant
Taurus	Tauri	Tau	Bull
Triangulum	Trianguli	Tri	Triangle
Ursa Major	Ursae Majoris	UMa	Great Bear
Ursa Minor	Ursae Minoris	UMi	Lesser Bear
Vela	Velorum	Vel	Sails
Virgo	Virginis	Vir	Virgin
Vulpecula	Vulpeculae	Vul	Fox

Some common asterisms

Belt of Orion	δ, ε and ζ Orionis
Big Dipper	α, β, γ, δ, ε, ζ and η Ursae Majoris
Circlet	γ, θ, ι, λ and κ Piscium
Guards (or Guardians)	β and γ Ursae Minoris
Head of Cetus	α, γ, ξ², μ and λ Ceti
Head of Draco	β, γ, ξ and ν Draconis
Head of Hydra	δ, ε, ζ, η, ρ and σ Hydrae
Keystone	ε, ζ, η and π Herculis
Kids	ε, ζ and η Aurigae
Little Dipper	β, γ, η, ζ, ε, δ and α Ursae Minoris
Lozenge	= Head of Draco
Milk Dipper	ζ, γ, σ, φ and λ Sagittarii
Plough or Big Dipper	α, β, γ, δ, ε, ζ and η Ursae Majoris
Pointers	α and β Ursae Majoris
Sickle	α, η, γ, ζ, μ and ε Leonis
Square of Pegasus	α, β and γ Pegasi with α Andromedae
Sword of Orion	θ and ι Orionis
Teapot	γ, ε, δ, λ, φ, σ, τ and ζ Sagittarii
Wain (or Charles' Wain)	= Plough
Water Jar	γ, η, κ and ζ Aquarii
Y of Aquarius	= Water Jar

Acknowledgements

Damian Peach, Hamble, Hants.: p.12 (Comet Lovejoy)

Sjbmgrtl: p.13 (Comet McNaught)

[https://commons.wikimedia.org/wiki/File.Sat_comet_WEB.jpg]

peresanz/Shutterstock: p.21 (Orion)

Steve Edberg, La Cañada, California: all other constellation photographs

Denis Buczynski, Portmahomack, Ross-shire, p.23 (Quadrantid fireball); p.51 (noctilucent clouds)

Jens Hackmann: p.63 (Perseid meteor)

Ken Sperber, California: p.69 (Double Cluster)

Steve Edberg, La Cañada, California: all other constellation photographs

Specialist editorial support was provided by Gregory Brown, Astronomy Education Officer at the Royal Observatory Greenwich, part of Royal Museums Greenwich.

Further Information

Books

Bone, Neil (1993), *Observer's Handbook: Meteors*, George Philip, London & Sky Publ. Corp., Cambridge, Mass.

Cook, J., ed. (1999), *The Hatfield Photographic Lunar Atlas*, Springer-Verlag, New York

Dunlop, Storm (1999), *Wild Guide to the Night Sky*, HarperCollins, London

Dunlop, Storm (2012), *Practical Astronomy*, 3rd edn, Philip's, London

Dunlop, Storm, Rükl, Antonin & Tirion, Wil (2005), *Collins Atlas of the Night Sky*, HarperCollins, London

Heifetz, Milton & Tirion, Wil (2017), *A Walk through the Heavens: A Guide to Stars and Consellations and their Legends*, 4th edition, Cambridge University Press, Cambridge

O'Meara, Stephen J. (2008), *Observing the Night Sky with Binoculars*, Cambridge University Press, Cambridge

Ridpath, Ian, ed. (2004), *Norton's Star Atlas*, 20th edn, Pi Press, New York

Ridpath, Ian, ed. (2003), *Oxford Dictionary of Astronomy*, 2nd edn, Oxford University Press, Oxford

Ridpath, Ian & Tirion, Wil (2004), *Collins Gem - Stars*, HarperCollins, London

Ridpath, Ian & Tirion, Wil (2011), *Collins Pocket Guide Stars and Planets*, 4th edn, HarperCollins, London

Ridpath, Ian & Tirion, Wil (2012), *Monthly Sky Guide*, 9th edn, Cambridge University Press

Rükl, Antonín (1990), *Hamlyn Atlas of the Moon*, Hamlyn, London & Astro Media Inc., Milwaukee

Rükl, Antonín (2004), *Atlas of the Moon*, Sky Publishing Corp., Cambridge, Mass.

Scagell, Robin (2000), *Philip's Stargazing with a Telescope*, George Philip, London

Tirion, Wil (2011), *Cambridge Star Atlas*, 4th edn, Cambridge University Press, Cambridge

Tirion, Wil & Sinnott, Roger (1999), *Sky Atlas 2000.0*, 2nd edn, Sky Publishing Corp., Cambridge, Mass. & Cambridge University Press, Cambridge

Journals

Astronomy, Astro Media Corp., 21027 Crossroads Circle, P.O. Box 1612, Waukesha, WI 53187-1612 USA. http://www.astronomy.com

Astronomy Now, Pole Star Publications, PO Box 175, Tonbridge, Kent TN10 4QX UK. http://www.astronomynow.com

Sky at Night Magazine, BBC publications, London. http://skyatnightmagazine.com

Sky & Telescope, Sky Publishing Corp., Cambridge, MA 02138-1200 USA. http://www.skyandtelescope.com/

Societies

British Astronomical Association, Burlington House, Piccadilly, London W1J 0DU. http://www.britastro.org/
The principal British organization for amateur astronomers (with some professional members), particularly for those interested in carrying out observational programmes. Its membership is, however, worldwide. It publishes fully refereed, scientific papers and other material in its well-regarded journal.

Federation of Astronomical Societies, Secretary: Ken Sheldon, Whitehaven, Maytree Road, Lower Moor, Pershore, Worcs. WR10 2NY. http://www.fedastro.org.uk/fas/
An organization that is able to provide contact information for local astronomical societies in the United Kingdom.

Royal Astronomical Society, Burlington House, Piccadilly, London W1J 0BQ. http://www.ras.org.uk/
The premier astronomical society, with membership primarily drawn from professionals and experienced amateurs. It has an exceptional library and is a designated centre for the retention

of certain classes of astronomical data. Its publications are the standard medium for dissemination of astronomical research.

Society for Popular Astronomy, 36 Fairway, Keyworth, Nottingham NG12 5DU.
   http://www.popastro.com/
   A society for astronomical beginners of all ages, which concentrates on increasing members' understanding and enjoyment, but which does have some observational programmes. Its journal is entitled *Popular Astronomy*.

Software

Planetary, Stellar and Lunar Visibility, (Planetary and eclipse freeware): Alcyone Software, Germany.
   http://www.alcyone.de
Redshift, Redshift-Live. http://www.redshift-live.com/en/
Starry Night & Starry Night Pro, Sienna Software Inc., Toronto, Canada. http://www.starrynight.com

Internet sources

There are numerous sites with information about all aspects of astronomy, and all of those have numerous links. Although many amateur sites are excellent, treat any statements and data with caution. The sites listed below offer accurate information. Please note that the URLs may change. If so, use a good search engine, such as Google, to locate the information source.

Information

Astronomical data (inc. eclipses) HM Nautical Almanac Office: http://astro.ukho.gov.uk
Auroral information Michigan Tech: http://www.geo.mtu.edu/weather/aurora/
Comets JPL Solar System Dynamics: http://ssd.jpl.nasa.gov/
American Meteor Society: http://amsmeteors.org/
Deep-sky objects Saguaro Astronomy Club Database: http://www.virtualcolony.com/sac/
Eclipses: NASA Eclipse Page: http://eclipse.gsfc.nasa.gov/eclipse.html
Moon (inc. Atlas) Inconstant Moon: http://www.inconstantmoon.com/
Planets Planetary Fact Sheets: http://nssdc.gsfc.nasa.gov/planetary/planetfact.html
Satellites (inc. International Space Station)
   Heavens Above: http://www.heavens-above.com/
   Visual Satellite Observer: http://www.satobs.org/
Star Chart http://www.skyandtelescope.com/observing/interactive-sky-watching-tools/interactive-sky-chart/
What's Visible
   Skyhound: http://www.skyhound.com/sh/skyhound.html
   Skyview Cafe: http://www.skyviewcafe.com

Institutes and Organizations

European Space Agency: http://www.esa.int/
International Dark-Sky Association: http://www.darksky.org/
Jet Propulsion Laboratory: http://www.jpl.nasa.gov/
Lunar and Planetary Institute: http://www.lpi.usra.edu/
National Aeronautics and Space Administration: http://www.hq.nasa.gov/
Solar Data Analysis Center: http://umbra.gsfc.nasa.gov/
Space Telescope Science Institute: http://www.stsci.edu/

POLITICAL
LONDON

POLITICAL LONDON

A Guide to the Capital's
Political Sights

J B Seatrobe

First published in Great Britain 2000
Published by Politico's Publishing
8 Artillery Row
Westminster
London
SW1P 1RZ
Tel 020 7931 0090
Fax 020 7828 8111
Email publishing@politicos.co.uk
Website http://www.politicos.co.uk/publishing

A catalogue record of this book is available from the British Library.
ISBN 1 902301 52 8

Printed and bound in Great Britain by St. Edmundsbury Press.
Cover Design by Advantage

Contents

Preface

This book is dedicated to the political sights and sites of London. There seem to be books about all imaginable aspects of London, but not one devoted solely to the metropolis as a capital city, seat of government, home of parliaments, refuge for political exiles from all over the world, and stage for some of the greatest political dramas in history.

Whether you are a tourist, interested in the rich political life and history of this great city, or a true political anorak eager to stand on the site of great events or outside the places where the famous have lived and worked, *Political London* is the book for you.

Political London isn't an A–Z listing of places and streets. It's a guide presented in various ways and themes to provide you with a genuine feel of the political life and history of London.

Deep gratitude is owed to Shona and Keith Skakle, without whose valuable research assistance and advice this book would have been much the poorer. Grateful thanks also go to Geoff Lindsey for his input. The enthusiasm and guidance of the staff at Politico's more than made up for their tight publishing deadlines! John Simmons provided the 'Political eating, drinking and shopping' section and compiled the index.

Responsibility for any errors rests as always with the author, who welcomes any suggestions and additions for future editions of *Political London*.

JBS

March 2000

POLITICAL
HOT SPOTS

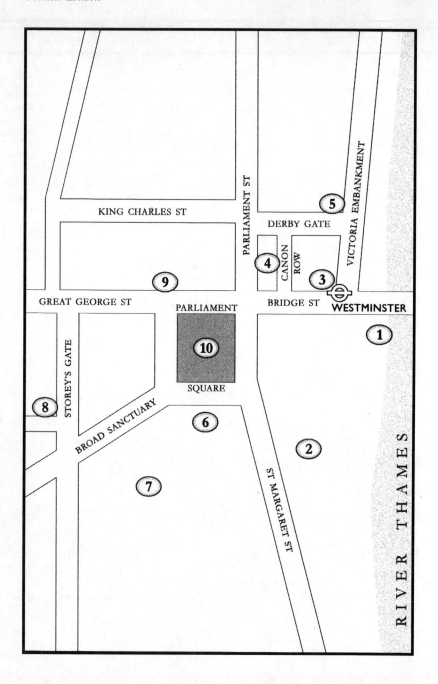

KING CHARLES ST

PARLIAMENT ST

DERBY GATE

VICTORIA EMBANKMENT

⑤

④

CANON ROW

③

⑨

GREAT GEORGE ST

BRIDGE ST

WESTMINSTER

①

PARLIAMENT

⑩

SQUARE

STOREY'S GATE

⑧

BROAD SANCTUARY

⑥

⑦

②

ST MARGARET ST

RIVER THAMES

Parliament Square

The area around Parliament Square brings together much of Britain's political and parliamentary history. The government buildings in Whitehall are to the north, the Houses of Parliament to the east and Westminster Abbey immediately to the south.

1. PALACE OF WESTMINSTER

A Gothic building facing the river, home to the two Houses of Parliament, the Commons to the north and the Lords to the south. If Parliament is sitting, there will be a 'Union Jack' flying over the Victoria Tower and, after dark, a light will be on at the top of the Clock Tower ('Big Ben'). Although access to the Palace is now generally strictly controlled, people can queue outside the St Stephen's entrance to listen to debates, and there are also guided tours by arrangement. Write to your MP at the House of Commons, London SW1A 0AA, or foreign visitors can apply to the Parliamentary Education Unit, Norman Shaw Building North, London SW1A 2TT (020 7219 4272).

Westminster Bridge or the riverbank opposite are the best places to see the front of

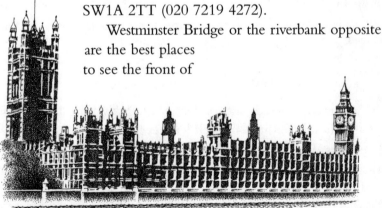

the Palace in all its glory, especially the riverfront terrace where MPs, peers and their guests congregate when the weather permits.

Parliamentarians can be seen coming and going through the large Carriage Gates into New Palace Yard, especially around 7pm and 10pm when there are likely to be votes. On Wednesday afternoons the Prime Minister is driven in on the way to the weekly half-hour of PM's Question Time. Most spectacular of all is the State Opening of Parliament, usually in November each year.

The House of Lords Record Office lives in the Victoria Tower, at the southern end of the Palace. This repository of the parliamentary archives includes original Acts of Parliament. The Office is open to the public, who can examine documents in the Search Room.

The Houses of Parliament have not been free of political violence. Apart from the 1605 Gunpowder Plot, the Prime Minister, Spencer Perceval, was shot dead in the lobby of the House of Commons on 11 May 1812. There was a bomb attack on Westminster Hall by Fenians in 1885 which caused a large hole in the floor, and the Conservative MP, Airey Neave, was killed by a car bomb on the ramp of the underground car park in New Palace Yard on 30 March 1979.

2. WESTMINSTER HALL

In medieval times, Royal or Great Councils were held in the Hall, although Parliament itself has never met there, as such. For centuries Westminster Hall was the seat of English justice with the law courts sited there. Other notable political or state occasions in the Hall include the trials of Sir Thomas More in 1535, Guy Fawkes in 1606, Thomas Wentworth, Earl of

Strafford, Charles I's Minister, in 1641, Charles I himself in January 1649, and Warren Hastings from 1788 to 1795. The last impeachment there was that of Viscount Melville in 1806, who was acquitted.

Westminster Hall has seen the lying in state of major figures, such as Gladstone in June 1898 and Churchill in January 1965. It was the scene of addresses to both Houses by heads of state – Presidents Lebrun and De Gaulle of France in 1939 and 1960 respectively, and by President Mandela of South Africa in July 1996.

The Hall is still used for important public events, such as the celebration of Winston Churchill's 80th birthday (including presentation of the notorious Sutherland portrait) on 30 November 1954; the 700th anniversary of Simon de Montfort's Parliament in 1965; the tercentenary of the Glorious Revolution in 1988, and the 50th anniversary of the United Nations in 1995.

3. PORTCULLIS HOUSE

The newest of the parliamentary buildings, constructed above the rebuilt Westminster tube station, it provides office space for over 200 MPs and their staff on seven floors, with committee rooms and other common facilities.

4. PARLIAMENT STREET BLOCK

A series of buildings which were incorporated as parliamentary accommodation in 1991. At the Bridge Street corner is the Parliamentary Bookshop. Between it and Canon Row was St Stephen's Tavern, a popular watering hole for politicians. The corner site had occupants from the National Labour Committee to the Department of Transport. Looking north up Parliament Street is the main parliamentary entrance, 1 Parliament Street, in what was formerly nos. 34–42. No. 38 was occupied by the notorious Maundy Gregory (see *Scandals*). Nos 2 and 3, formerly 43–44, were the residences for parliamentarians, offices for parliamentary clerks, and for several government bodies. No. 43 was for a time a Liberal Party office, and no. 44 was well-known as a post office for parliamentarians and tourists alike. At the north end of the block is 1 Derby Gate, formerly nos. 45–47 Parliament Street, now housing much of the House of Commons Library. No. 47 was for forty years the Whitehall Club, and from 1966 to 1972, the Welsh Office.

5. NORMAN SHAW BUILDINGS

These much-filmed buildings, designed by Richard Norman Shaw (1890) formerly housed the HQ of the Metropolitan Police, 'New Scotland Yard', after their move from the original Scotland Yard, further up Whitehall. Nowadays, as Norman Shaw North and Norman Shaw South, they are used as MPs' offices.

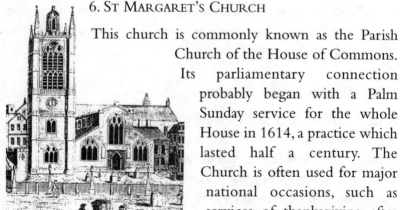

6. ST MARGARET'S CHURCH

This church is commonly known as the Parish Church of the House of Commons. Its parliamentary connection probably began with a Palm Sunday service for the whole House in 1614, a practice which lasted half a century. The Church is often used for major national occasions, such as services of thanksgiving after the two world wars in the twentieth century. A service is held at the beginning of each new parliament, and the church hosts numerous religious services for Members and their families, from weddings to memorial services.

7. WESTMINSTER ABBEY

In the Abbey are buried some of Britain's most illustrious political figures, including no less than eight Prime Ministers. Others include Wilberforce, Fox, Ernest Bevin and Beatrice and Sidney Webb. There are many statues and memorials to other political notables, including Asquith, Peel and Disraeli.

8. CENTRAL HALL, WESTMINSTER

Built early last century as the Methodist Church's HQ, it became a convenient location for public events and political meetings. Many famous political figures have spoken here, including Lloyd George (attempting one of his comebacks in 1935) and foreign leaders such as de Gaulle, Gandhi and Gorbachev. Labour and Conservative party conferences were held here during both world wars. It was also the venue for a special Labour conference on Common Market entry in July 1971.

Perhaps Central Hall's finest hour was in early 1946 with the inaugural meeting of the General Assembly of the United Nations. Ernest Bevin, the Foreign Secretary, had to persuade the Hall's trustees that the UN meeting should be held there. Even so, the availability of a bar for the delegates upset the Methodist Church. The General Assembly first met on 10 January 1946. The Labour Prime Minister, Clement Attlee, unveiled a commemorative plaque on 28 May 1946.

9. GREAT GEORGE STREET

Its close proximity to Parliament made this a popular residential street for MPs and peers. George Grenville lived there; John Wilkes lived at no. 13 between 1757 and 1763, and Sir Robert Peel at no. 36 in 1813. The architect of the new Houses of Parliament, Sir Charles Barry, lived at no. 32 between 1859 and 1870. The north side of the street disappeared in 1910 when what is now the vast Treasury building was constructed.

10. PARLIAMENT SQUARE

The Square is filled with political statues, from that of Churchill (unveiled by Lady Spencer Churchill, 1973) to those of the Victorian premiers Peel (1850), Derby (1874), Palmerston (1876) and Disraeli (1883). Next to the Lincoln statue is Canning Green, named after the Tory PM, George Canning. Canning's statue (erected in 1832, at a cost of £7,000) originally stood nearer Westminster Hall, but was moved in 1867 when alterations were made to the layout of the area. Although there are strict laws about assemblies close to Parliament, the Square often hosts protest demonstrations by those keen to publicise their case with parliamentarians.

Whitehall

Whitehall is known as the centre of British government, even though many government offices have been dispersed. It's not just Downing Street, the Cenotaph, the state occasions and the massive piles housing the Ministry of Defence, the Treasury and the Foreign Office. It's also its long history,

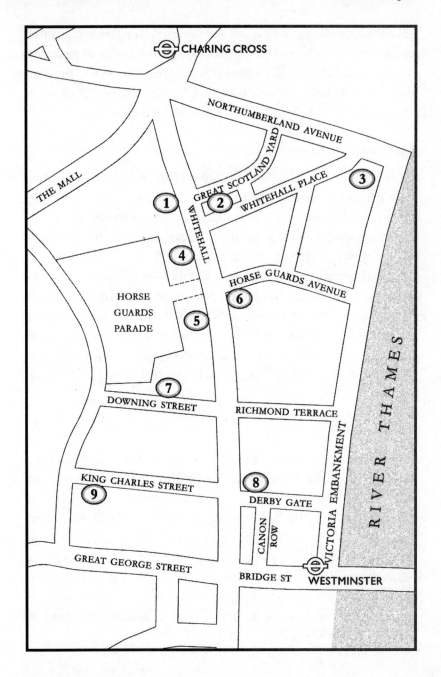

when the sprawling, long-gone Whitehall Palace (covering much of the site) was the seat of English government. The little streets off Whitehall (some, like historic King Street, now lost beneath more recent development) provided homes and places of work and entertainment for all involved in matters of state.

1. ADMIRALTY HOUSE

Its apartments were used by First Lords of the Admiralty (including Balfour and Churchill) until 1960. It became the temporary residence of Harold Macmillan during the renovation of No. 10 Downing Street, and is used as a residence for the Defence Secretary and other ministers. In the 1990s the Prime Minister operated from here during repairs at No 10 following terrorist attacks. This had disastrous consequences during 'Black Wednesday', 16 September 1992, when key members of the government were forced to cope with the ERM financial crisis with virtually no means of communication. Apparently the Chief Whip was sent to fetch a radio, so that they could follow events!

2. WHITEHALL PLACE

It is flanked at the Whitehall end by the Old War Office Building and offices of the Ministry of Agriculture, Fisheries and Food. No. 4 was a private house, which became the offices of the commissioners of Peel's Metropolitan Police in 1829. The back, used as a police station, opened on to a courtyard, known as 'Scotland Yard', because it was on a site of a house used by Scottish monarchs visiting London. Hence the name by which the Metropolitan Police HQ has become immortalised, even when it moved in 1890 to the Victoria Embankment, and in 1967 to Broadway, SW1.

3. NATIONAL LIBERAL CLUB

No. 1 Whitehall Place is better known as the National Liberal Club, formed in 1882 with Gladstone as its first President. In 1916 the Club was pressed into war duties. The Liberal Party and the European Movement have had their headquarters on the premises.

4. 36 WHITEHALL

This modest building is a good example of the political history of Whitehall. In the 17th century the Office of the Paymaster of the Forces was on the site. The present building was constructed in the 1730s, with additions in 1806. Following restoration (after wartime bomb damage) involving new internal offices within the retained facades, the building was occupied by the Parliamentary Counsel Office. There is a reconstruction of the historic Paymaster General's room on the ground floor.

5. DOVER HOUSE

Built in the 1750s, it became the French Ambassador's home, and later York House (1788-92 when the Duke of York lived there). When the Duke exchanged it for Lord Melbourne's Piccadilly house, it was renamed Melbourne House between 1793-1830, and then became Dover House. It was offered to Gladstone, when PM, as an alternative to Downing Street, but he refused because he feared that its grandeur compared to No. 10 would require him to hold more social events, destroying his privacy. It was occupied by the Scottish Office in 1885, remaining in Scottish hands thereafter, even after devolution. Across Whitehall is Gwydyr House, the Welsh Office.

6. BANQUETING HOUSE

The last surviving remnant of Whitehall Palace, it dates from the early 1620s. As the plaque on a bust of Charles I above the entrance notes, it was here that the unfortunate monarch was beheaded on 30 January 1649. It is used for major occasions, such as the lunch on 20 November 1997, where the Prime Minister, Tony Blair, made a speech celebrating the golden wedding anniversary of the Queen and the Duke of Edinburgh. It is the only major building in Whitehall open to the public, and there is a fascinating picture in the entrance showing a proposed seating plan, dated 7 November 1936, for the possible wedding and coronation of Edward VIII.

7. DOWNING STREET

A short, narrow street, it is probably the best-known road in Britain, with No. 10, No. 11 (the official residence of the Chancellor of the Exchequer) and No. 12 (the office of the Government Chief Whip). No. 10, with one of the world's most recognisable front doors (actually there are two, to allow for repairs and renovation), is the Prime Minister's official residence, by virtue of being First Lord of the Treasury. The brass plaque on the door proclaims 'First

Lord of the Treasury', and when various PMs, such as Churchill, asked that the plaque be removed, they were told that they would have to pay rent for living there as Prime Minister.

Robert Walpole, Britain's first Prime Minister, was offered No. 10 by George II in 1732, and moved in three years later. Since then, most premiers have lived there when in office. No. 10 is not as small as it may appear from its Downing Street frontage. It has over 160 rooms, and it is actually two linked houses, one (Albemarle House, later Bothmar House) being a large three-storey property facing Horse Guards Parade.

The street was so named by George Downing, who had a varied career, from public official to MP during the 17th century. He acquired property west of Whitehall, which he demolished and built the houses in the street. Of the original development, only Nos. 10 and 11 remain; No. 12 was totally remodelled in the 1960s. The street used to be accessible to the public, and was regularly filled with tourists, some perhaps emulating the eight-year-old Harold Wilson, who, in 1924, had his photo taken outside No. 10. Because of the terrorist threat, the street was closed to the public in the 1980s, a security barrier was put up in 1981, and a set of gates were erected in 1989-90, which remain to this day.

8. THE RED LION PUBLIC HOUSE

This is one of the surviving reminders of the dozens of drinking places which were dotted around Whitehall. It is on the site of a medieval tavern. The present pub was rebuilt in 1900, and is a regular haunt of parliamentarians.

9. CABINET WAR ROOMS

The fortified shelter for Winston Churchill, his Cabinet and senior military and intelligence advisers during World War II. The various rooms where they lived and worked have been open to the public for a number of years.

Smith Square

The area around Smith Square is full of political interest, and, in conjunction with Parliament itself, helps make 'Westminster' the generic name for parliamentary and party political activity. Conservative Central Office has been in the Square since 1958; Labour's HQ was Transport House, 1928-1980, and the Liberals' HQ was in the Square for a time in the 1960s.

1. SMITH SQUARE

It's not just Transport House and Conservative Central Office (at no. 32) which make Smith Square 'political'. The leading Conservative, Rab Butler lived for a time at no. 3, west of Lord North Street, until it was bombed in 1940, and the fascist leader, Sir Oswald Mosley lived at nos. 8-9, to the east of the street. Many connected with local political life dine in the Footstool restaurant in St John's Church in the centre of the Square.

2. TRANSPORT HOUSE

Designed as the HQ of the newly formed Transport and General Workers Union, it was intentionally located near Parliament and Government. In 1925, it was agreed that the TUC and the

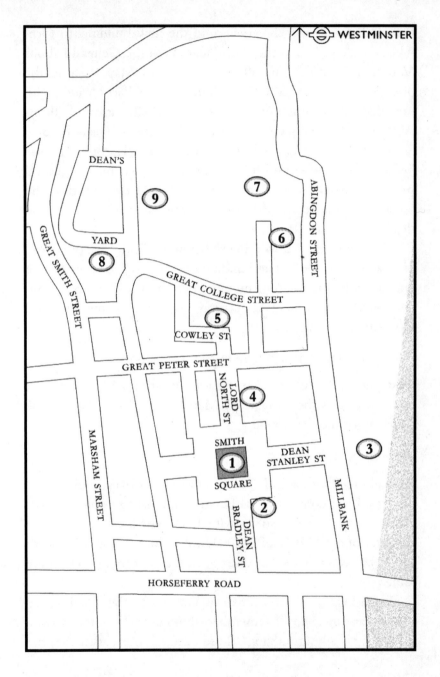

Labour Party would also move into the new building, to form what was described as the "Headquarters of the General Labour Movement". It was formally opened on 15 May 1928, in the presence of Ramsay MacDonald, the Labour Party leader, who unveiled a commemorative tablet on the ground floor. Although the main entrance was originally in Dean Bradley Street, it was as a part of Smith Square that Transport House became best known. With the HQs of the two major political parties practically next door to each other, at election time the Square was the scene of joyous triumph, as well as defeat and dejection.

The TUC moved to a new HQ in 1957, the Labour Party did likewise in 1980. The building's history as Transport House ended just before the millennium when the TGWU moved out. The Local Government Association moved in, renaming it Local Government House.

3. The Buxton Memorial Drinking Fountain

Designed and built by Mr Charles Buxton MP in 1865, it commemorates the struggle by Members of Parliament for the abolition of the slave trade.

4. Lord North Street

A small street with a lively political history. Virtually every house has a political connection, although, contrary to popular belief, the street has nothing to do with Lord North, the Tory Prime Minister of the 1770s and 1780s. Until 1936, it was simply North Street, but was renamed to distinguish it from other streets of the same name.

It has been popular with Conservatives, as Alan Clark noted in his diary in 1983: "I remember thinking these houses were a bit poky, blackly crumbling .. I see now, of course, that they are the choicest thing you can have if you are a Tory MP". Anthony Eden lived at no.2 in 1924; Sir Edward du Cann at no. 5 from 1965 to 1970 and no. 19 from 1970 to 1978; Brendan Bracken (1931-58) and, more recently, Sir Nigel Fisher lived at no. 10; Lord Dilhorne at no. 11, and Teresa Gorman at no. 14. No. 8 has been a hotbed of activity. During Bracken's residency in the 1930s many of the Party's anti-appeasers rallied round Churchill there, and when Jonathan Aitken lived there his home was the venue for meetings of the Conservative Philosophy Group, with speakers such as Margaret Thatcher and Richard Nixon.

The most famous non-Conservative resident was the Labour PM, Harold Wilson, who lived for at no.5 for six years. He moved there after his election defeat in 1970 and left after his surprise resignation as Prime Minister in 1976.

5. 4 COWLEY STREET

In the 1980s the Social Democratic Party was based here. Following the SDP's merger with the Liberal Party in 1988, it became the HQ of the Liberal Democrats.

6. COLLEGE GREEN

This area of grass across from the Houses of Parliament has become famous as the place for TV interviews with MPs and ministers. It came to the forefront of the public's mind during the high drama of the Conservative Party leadership election in November 1990, which led to the toppling of Margaret Thatcher in favour of John Major.

7. JEWEL TOWER

Built in the 14th century, it was for over two centuries from 1621 the Parliament Office, housing Parliament's records. After

70 years as a part of the Weights and Measures Office, it underwent major renovation work, and now hosts a fascinating public exhibition on the history and role of Parliament.

8. CHURCH HOUSE, DEAN'S YARD

Having survived a serious air raid in October 1940, Church House was seen as a possible alternative venue for Parliament to meet. Parliament met there during three stages of the War, in late 1940 (from 7 November, with the new session opened by the King on the 21st); spring 1941 (caused by the direct hit on the Commons on 10-11 May) and summer 1944. The Commons met in the Hoare Memorial Hall (a plaque marking this was unveiled by Attlee and Churchill on 28 April 1948), and the Lords in the Convocation Hall. Church House, jokingly known as the 'Churchill Club' during these periods, was the scene of many dramatic wartime parliamentary occasions.

At the end of the War it was the site of various United Nations commissions. The War Crimes Commission met in June 1945, and the preparatory commissions of the UN itself met there in late 1945. The Security Council met in Bishop Partridge Hall in January 1946. These events are commemorated in a stone in the Entrance Porch, unveiled by the then UN Secretary General, Pérez de Cuellar, in 1986.

Because of its location, Church House remained a venue for political and state meetings of all kinds. Two prime ministers of the 1950s, Anthony Eden and Harold Macmillan, were formally elected as party leader by Conservative MPs there, in April 1955 and January 1957 respectively. Many high-profile public inquiries and tribunals have been held there, such as the Bank Rate Leak Tribunal in 1957 and the Scarman Inquiry into the 1981 Brixton riots. Prime Minister Tony Blair met his huge

Parliamentary Labour Party there following his landslide general election victory in May 1997.

9. WESTMINSTER SCHOOL

This school off Dean's Yard has had many famous political pupils, including some future prime ministers - Henry Pelham, the Duke of Newcastle, the Marquess of Rockingham, the Duke of Portland and Lord John Russell.

St James's

Tucked between seats of government and monarchy, with St James's Square at its centre, this small rectangle of central London was, and to some extent remains, the heart of political society and clubland.

1. ST JAMES'S SQUARE

No. 4 was Nancy Astor's home, the first woman to sit in the House of Commons. No. 10 was, appropriately, home to three prime ministers, Chatham, Derby and Gladstone (commemorated by a plaque). As Chatham House (joined with no. 9), it is the office of the Royal Institute of International Affairs. From no. 31, Eisenhower directed the Allied Invasion of Europe in 1944. Lord Derby lived for the last fifteen years of his life at no. 33.

The Square has other Prime Ministerial connections: Walpole lived there during his premiership, and his little-known successor, the Earl of Wilmington, died there. The Duke of Grafton was married at no. 13 in 1756, and Lord Grenville lived at no. 17.

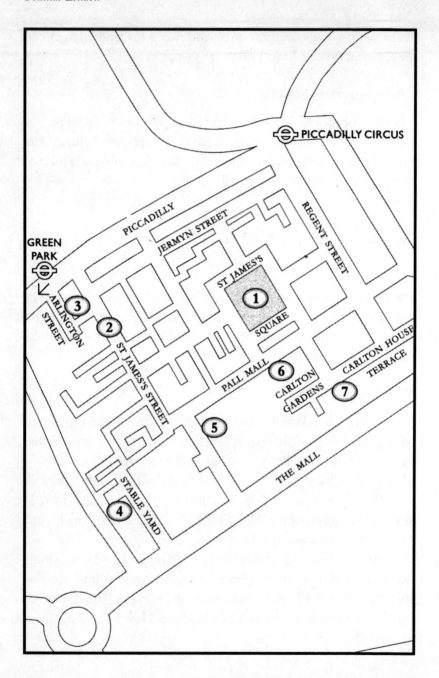

One notorious episode in the 1980s took place outside the then Libyan Embassy, the fatal shooting of PC Yvonne Fletcher during disturbances on 17 April 1984.

2. St James's Street

This street still houses many political clubs. It began with establishments such as White's Chocolate House, a gambling den and Tory club, established in 1693, and the Cocoa Tree Tavern, a Tory chocolate house in 1698.

Boodle's, at no. 28, was established in Pall Mall in 1762 as a political club of Lord Shelburne's supporters, but the political connection soon lapsed. White's, at no. 37, had many notable political members, including lots of prime ministers, but became primarily non-political after the formation of the Carlton Club in 1832. Charles James Fox lived in the street, and was a member of Brooks's (no.60). When his debts threatened to overwhelm him, he was rescued by some of his fellow members.

Probably the best known of the clubs is the Carlton, the main Tory establishment. It grew out of unofficial attempts at party organisation by the 'Charles Street Gang', operating out of the home of the Mayfair home of Joseph Planta. It gained political immortality when it was the venue of the meeting of Tory MPs in October 1922 which led to the removal of the Lloyd George coalition government, to be replaced by a Conservative administration under Bonar Law. This meeting was the genesis of the Conservative backbench parliamentary '1922 Committee'.

3. Arlington Street

This small street has played host to many political figures. Britain's first prime minister, Sir Robert Walpole, who lived at no. 17 from 1716 to 1731, died in 1745 at no.5 (marked by a

GLC plaque). A plaque at no. 9 records that Charles James Fox lived there from 1804-6. Others who lived in the street included the Salisburys, and the 18th century politicians, Lord Carteret and William Pulteney.

No.22 was built in 1743 for long-time PM of the mid-18th century, Henry Pelham, who died there in 1754. The house was later known as Hamilton House and, from 1880, as Wimborne House. In 1926 it was the scene of talks between the coal owners and trade union leaders which led to the ending of the General Strike.

4. LANCASTER HOUSE

This was formerly known as Godolphin House (where Charles James Fox lived shortly before his death), York House (when occupied by the Duke of York) and Stafford House (where the social reformer Shaftesbury and William Garrison, the US slavery abolitionist spoke). It became Lancaster House in 1913 and was given to the nation the following year. It is frequently used for major state and diplomatic occasions. Churchill as PM hosted a banquet for the Queen shortly after her coronation in 1953, and Edward Heath announced Stage II of his government's prices & incomes policy at a press conference here in January 1973. It was also the site of major conferences on Kenya in the early 1960s; Rhodesia/Zimbabwe (1979-80); Bosnia (July and December 1995); Nazi gold in December 1997, and on humanitarian aid to Iraq in April 1998.

5. MARLBOROUGH HOUSE

This was the 'court' of the Prince of Wales after the death of Prince Albert in the late 19th century. It was the site of Marlborough House Balls, especially the fancy dress ball on 22

July 1874, which saw the Prince dressed as Charles I. Gladstone was allowed to appear in uniform instead of fancy dress after he had been dining and speaking at Mansion House. It now houses the Commonwealth Secretariat.

6. PALL MALL

Like St James's Street, home to many clubs. One of the more famous was the Reform at no. 104, which began in 1836 as a meeting place for radicals, and was long associated with leading Liberal politicians from Gladstone and Palmerston in the 19th century, to Lloyd George and Churchill in the 20th. Sir Henry Campbell-Bannerman was elected Liberal leader there in 1899. Other clubs include the Travellers' at no. 106 (from which Lord Rosebery and Lord Randolph Churchill were blackballed), and the Athenaeum at no. 107.

7. THE CARLTONS

Carlton Gardens and Carlton House Terrace are rich in political connections. Gladstone lived in nos. 13 (1839–48) and 11 (1856–75) Carlton House Terrace, broken by 8 years at 6 Carlton Gardens (1848–56).

No. 1 Carlton Gardens was also the residence of Viscount Goderich (the Tory PM) and, when in exile, Louis Napoleon. It is the official residence of the Foreign Secretary. No. 3 was said to be the base for MI6's recruitment. No 4 was home to two prime ministers, Balfour and Palmerston, the latter noted by a plaque. De Gaulle lived there from 1940 to 1943, when it was the HQ of Free French forces. There is a Cross of Lorraine memorial and statue of de Gaulle nearby. Margaret and Denis Thatcher had their wedding reception on 13 December 1951 at no. 5.

Lord Curzon lived at one Carlton House Terrace, and died there, and there is a statue of him opposite at the corner of Carlton Gardens. The Conservative PM, Lord Derby, lived at no. 11 (there is a plaque, shared with Gladstone).

Berkeley Square

Not surprisingly, such a fashionable square has been home to many political leaders over the decades.

1. BERKELEY SQUARE

Lord Rosebery lived at no. 2 and at no. 38 (the latter is now Berger House); Lewis Harcourt at no. 14 and Pitt the Younger at no. 47 (now HSBC Bank). The latter belonged to Pitt's elder brother, John, 2nd Earl of Chatham. Earl Grey lived at no. 48 (now the ScotiaBank), which he rented out between 1830 and 1834 to Henry, Lord Brougham, the Lord Chancellor, who apparently left the house in a terrible mess. Canning lived at no. 50 (said to be the haunted house of the Square) in 1806-7, marked by a GLC plaque.

2. BRUTON STREET

George Canning lived at no. 24. At no. 17, across the road on the south side, on 21 April 1926 the present Queen Elizabeth II was born. There is a private plaque to mark this event.

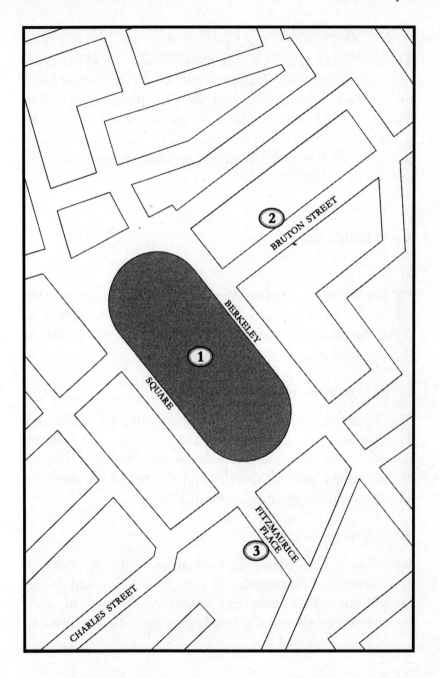

3. LANSDOWNE HOUSE

The Lansdowne Club at no. 9 Fitzmaurice Place is all that is left of Lansdowne House, which was built in 1761 by Robert Adam for the Earl of Bute. Bute sold the property for £22,500 to another PM, Shelburne, who took up residence in 1768. The treaty of American Independence was drafted in the Round Room in 1783. Harold Macmillan's wedding reception was held here on 21 April 1920.

Grosvenor Square

This Square has been the American heart of London since the 1780s. Its transatlantic connections earned it the title of 'Little America', and during the last War, it was known as 'Eisenhowerplatz'.

1. THE UNITED STATES EMBASSY

The Square is dominated by the monumental pile of the US Embassy, with its huge eagle, on the west side. The Embassy moved here in 1938, and the modern edifice (much of which is actually below ground) dates from 1960. It was the focus for anti-Vietnam War demonstrations in 1968.

2. GROSVENOR SQUARE

John Adams (2nd president) lived at no. 9, in the 1780s. A commemorative plaque states, in part: 'John Adams and Abigail his wife through character and personality did much to create understanding between the two English-speaking countries. In

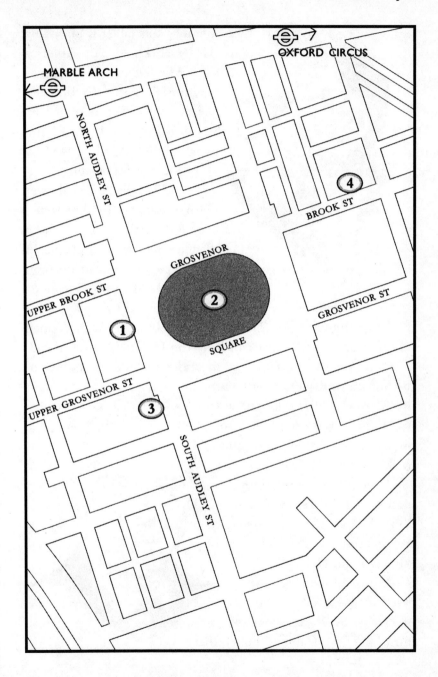

their memory this tablet is placed by the Colonial Dames of America 1933.' The Square's most famous memorial is that to Franklin Delano Roosevelt, designed by Sir William Reid Dick and unveiled in April 1948 by Eleanor Roosevelt.

Dwight D. Eisenhower (34th president) headquartered at no. 20, which is marked by a US Defense Department plaque. In 1944 he directed the Allied Invasion of Europe from 31 St James's Square, St James's SW1, which is also marked by a plaque. Ike lived in Kingston upon Thames, marked by a borough council plaque nearby at the corner of Kingston Hill and Warren Road.

Non-American political connections include the homes in the Square of the Duke of Grafton, the Marquess of Rockingham at no. 4, and, at the corner with South Audley Street, of John Wilkes. At no. 41, lived (and, in 1792, died) Lord North. North used to let his house out on short leases, often to newly-weds, earning it the nickname of 'Honeymoon Hall'.

At no. 39 (later renum-bered as 44) lived the Earl of Harrowby. It was here on 21 June 1815, at a Cabinet dinner, that the first news arrived of Wellington's victory over Napoleon at Waterloo. The house was known thereafter as 'Waterloo House', and was, five years later, intended to be the site of the assassination of the Cabinet at another dinner, by the Cato Street conspirators (see *Plots*).

3. SOUTH AUDLEY STREET

Lord John Russell lived at no. 66, and the Earl of Bute died at Bute House, no. 75. Palmerston lived on the other side of the Square, in North Audley Street.

4. BROOK STREET

William Pitt, Earl of Chatham, lived at no. 68. At no. 43 was a Conservative club, the Bath Club. It began in Dover Street in 1894, moved to new premises in St James's Street after the original site burned down during the Second World War, and merged with the Conservative Club in 1950. The club moved to Brook Street in 1959, though the site now houses a private bank. On the other side of the Square its sister street, Upper Brook Street, was the home of Baldwin and of George Grenville.

Hanover Square

L ike Berkeley Square, Hanover Square was a fashionable centre for the political elite, especially the prime ministers of the 18th and 19th centuries.

1. HANOVER SQUARE

Palmerston owned property here, Earl Grey lived in the Square and Lord Grenville married Anne Pitt, daughter of Lord

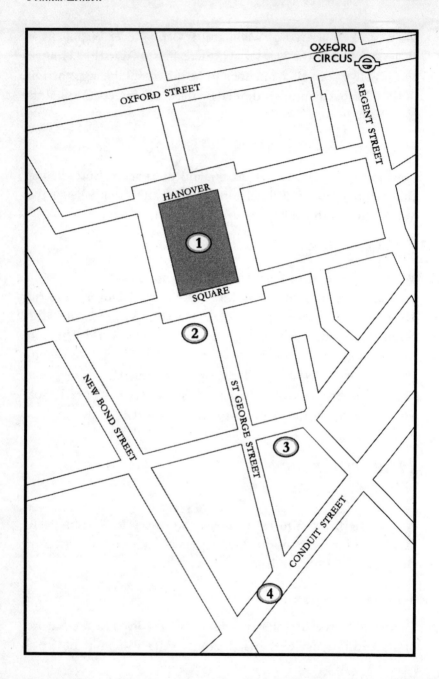

Camelford, here in 1792. One of the Marquess of Rockingham's close political allies, the Yorkshire MP Sir George Savile, lived in what was described as 'one of the most expensive houses' in the Square from 1755 to 1764. Talleyrand lived at no. 21 in the 1830s. At the south end of the Square is the large bronze statue of William Pitt the Younger, by Sir Francis Legatt Chantrey. It was erected in 1831, and survived attempts by Reform Bill supporters to pull it down on its unveiling day.

2. ST GEORGE STREET

Known as George Street until 1938, two Lord Chancellors, Cowper and Lyndhurst, lived here, as did RB Sheridan, who was an MP as well as a noted dramatist. Disraeli lived for a time at Edward's Hotel.

3. ST GEORGE'S CHURCH

This famous church has seen many political marriages, including those of Castlereagh (1794); Addington (1823); Grey (1805); Russell (1835); Palmerston (1839); Disraeli (1839) and Asquith (1894). When Asquith married Margot Tennant on 10 May 1894, four Premiers signed the register – Gladstone, Rosebery, Balfour, Asquith – and a fifth, Campbell-Bannerman, was in the congregation. Asquith lived at 20 Cavendish Square for a time, just across Oxford Street from Hanover Square.

4. CONDUIT STREET

Charles James Fox was born here in January 1749. George Canning lived at no. 37, and Lord Liverpool also lived in the Street.

Also in and around London

There are many other areas of London which are mini-Hot Spots in their own right, and well worth exploring. Many are covered in the following sections of this book within their particular themes. Here are some of the more interesting:

BELGRAVIA

This large sweep of SW1 contains several hot spots, primarily clustered around Eaton and Belgrave Squares, belying Disraeli's caustic comment that 'The Belgrave District is as monotonous as Marylebone, and is so contrived as to be at the same time insipid and tawdry'. In Eaton Square itself, lived such diverse characters as Metternich (at no. 44, in 1848, where Wellington was a frequent caller), Lord John Russell (at no. 48, in 1858), Stanley Baldwin (at no. 93, from 1913 to 1923), Lord Halifax, who lost to Churchill in the battle to succeed Chamberlain as PM in May 1940 (at no. 86) and the flamboyant Tory MP, Bob Boothby (at no. 1, from 1946 to 1986).

In adjacent Eaton Place, lived Sir Edward Carson (at no. 5), Sir John Lubbock, first Baron Avebury, promoter of the Bank Holidays Act 1871 (at no. 29). At no. 16 lived the MP William Ewart, the Victorian MP who was not only responsible for the first Public Libraries Act in 1850, but, by being the first advocate of a scheme to commemorate the houses of the famous in London, was invaluable in identifying the political sites of the capital for future generations, such as readers of this book.

South of Eaton Square lies South Eaton Place home to luminaries such as Robert Cecil, Viscount Cecil of Chelwood, a founder of the League of Nations (at no 16, from 1923 to 1958), the noted peace campaigner, Philip Noel Baker, and the controversial Conservative MP, Enoch Powell.

To the north, before reaching Hyde Park, is the Belgrave Square area. A landmark is the monument to Simon Bolivar, unveiled in 1974 by James Callaghan, then Foreign Secretary, in the south-east corner of the Square, opposite Upper Belgrave Street. An influential diarist, Henry 'Chips' Channon, held court at no. 5 from 1935 until his death in 1958. His entry for the first of February 1943 characteristically claimed that '5 Belgrave Square has played a considerable role in politics and society, and since the war has been, if it was not already so before, the centre of London ... no true chronicler of the time could fail to record its glories and its influence.' Walter Bagehot, editor of the *Economist* and a hugely influential writer on the British constitution, lived at 12 Upper Belgrave Street from 1860 to 1877.

THAMES RIVERSIDE

South of the Palace of Westminster the riverside area is full of political interest, especially in Millbank. At no. 4 is a media centre from which TV and radio companies broadcast their coverage of political and parliamentary affairs. You can often see famous politicians going to and from interviews. 7 Millbank is currently used as Parliamentary offices, and many government and public bodies are based in the area. Further down the road, just past Lambeth Bridge, is Thames House, an anonymous looking building which is the HQ of MI5. Just across the river at Vauxhall Cross, conveniently, the ostentatious yellow and green building is the headquarters of MI6. Also on Millbank is Millbank Tower, where the Labour Party has its head office.

Beyond Pimlico is the Chelsea Embankment, where there is a statue of Sir Thomas More. The noted Victorian Liberal politician, George Robinson, Marquess of Ripon lived at no. 9. The Chelsea Hospital, in adjacent Royal Hospital Road, has several prime ministerial connections. Wellington lay in state there in the Great Hall in 1852, ironically 44 years after facing a humiliating court of enquiry at the same site following the infamous Convention of Cintra during the Peninsular War. The Earl of Wilmington, Walpole's successor, was a Treasurer of the Hospital and the Earl of Bute was a Commissioner. Carrying on west along the Chelsea riverside, Cheyne Walk has been home to many notables, such as Sylvia Pankhurst, Lloyd George and, more recently, the Tory minister, Paul Channon. In adjacent Cheyne Row was no. 5, the residence of Thomas Carlyle, the 'Sage of Chelsea'.

BLOOMSBURY

Bloomsbury Square has a statue of Charles James Fox, and no. 6 was the home of Isaac D'Israeli and his more famous son. 140 Gower Street was a former HQ of MI5. The Georgian PM, Shelburne, was married in 1779 at St George's, Bloomsbury Way. The suffragist, Lady Jane Strachey, died in 1928 in Gordon Square.

COVENT GARDEN

King Street became noted as the site of the HQ of the Communist Party, and, at no. 38, is the Africa Centre, opened by President Kaunda of Zambia in November 1964. At 40 Maiden Lane lived Andrew Marvell, who was not only a literary figure, but was also MP for Hull. Georgiana, Duchess of Devonshire, took lodgings in Henrietta Street when campaigning for Fox during the famous 1784 Westminster election. There was a noted rally against Irish Home Rule at the Royal Opera House, Floral Street, on 16 May 1886.

THE CITY

Being the ancient heart of London, the 'square mile' and its environs have a long and varied political history. There are many sites connected with the Lord Mayor, Richard Whittington, of pantomime fame. He lived on a house on the site of 19 College Hill, and founded and was buried in St Michael Paternoster Royal Church in that street. Sir Francis Walsingham, Elizabeth I's minister, died in Seething Lane, off Great Tower Street in 1590. Wellington and Peel were original members of the City of London Club at 19 Old Broad Street. In Salisbury Square there is a memorial to Robert Waithman MP, a Lord Mayor and supporter of parliamentary reform, erected in 1833. Oliver Cromwell was married at St Giles, Cripplegate in 1620. Disraeli was articled to a firm of solicitors, Swains in 1821-24 at 6 Frederick's Place, Old Jewry, and Asquith was educated at the City of London School, Milk Street, off Cheapside. Fleet Street was not only synonymous with the print media, it also has a bronze bust of the Parnellite MP and journalist, TP O'Connor, and at no. 69 there is a plaque to John Bright and Richard Cobden, leading lights in the Anti-Corn Law League in the 1840s, which was based here.

EAST END

Sidney Street E1 was the site of a famous police siege of anarchists in January 1911. The celebrated 1888 'match-girls' strike' took place at the Bryant & May factory in Fairfield Road, Bow (marked by a plaque at the Fairfield Works). Attlee lived at Commercial Road, Limehouse, and in residences at two East London social organisations, Haileybury House, Durham Road, Stepney and Toynbee Hall in Whitechapel. The local and national Labour leader, George Lansbury, lived in a house on the site of

39 Bow Road E3 for 23 years, and the long-time Labour politician, Manny Shinwell, was born in 1884 at Brune House, Toynbee Street, Spitalfields. In Bow Churchyard, Bow Road, there is a bronze statue of Gladstone, by Albert Bruce-Joy, 1882, a gift of Theodore Bryant (of match company), to commemorate WEG's 50 years as an MP.

HAMPSTEAD

The area of north-west London in and around Hampstead has long associations with the left/progressive end of politics (hence the term 'Hampstead liberal'). It was home to Labour leaders such as Ramsay MacDonald (9 Howitt Road, Belsize Park, and then 103 Frognal), Hugh Gaitskell (18 Frognal Gardens), Harold Wilson (10 and 12 Southway, Hampstead Garden Suburb) and Michael Foot (Pilgrim's Lane). Beatrice and Sidney Webb lived at 10 Fitzjohn's Avenue and briefly at 10 Netherall Gardens, and George Orwell lived at 77 Parliament Hill in 1945, and worked at Westrope's Bookshop (on the site of the later Prompt Corner Café), South End Green in 1934. Asquith lived in Eton House, John Street (now Keats Grove), and at 4 Maresfield Gardens. The unfortunate Spencer Perceval lived at Belsize House, and Perceval Avenue is named after him. Parliament Hill is said to be so called because of the legend that Guy Fawkes and his co-plotters planned to watch the explosion of the Houses of Parliament from there. The notorious Kit Cat Club met at the Upper Flask Tavern, off East Heath Road.

POLITICAL
LANDSCAPES

Plots

The dark streets and alleyways of London have been home to innumerable plots, and conspirators have come from far and wide to undermine the political status quo.

SPA FIELDS TO CATO STREET

The early 19th century was a period of domestic unrest in London. A mass meeting was organised for Spa Fields, Clerkenwell, EC1 on 2 December 1816. Henry 'Orator' Hunt spoke to a large crowd at the nearby Merlin's Cave Tavern. When the authorities dispersed the crowd, some of the demonstrators tried to march on the City.

One Spa Fields organiser, Arthur Thistlewood, pursued more militant means. On 22 February 1820 he was told of a planned Cabinet dinner the following evening at the home of the Lord President of the Council, the Earl of Harrowby at no. 39 (later 44) Grosvenor Square, W1. He planned to murder them, and topple the government. A scheme was quickly concocted with fellow conspirators in a hayloft over a stable at no. 6 (later 1a) Cato Street, W1. Two sacks were brought to carry off the heads of two senior ministers, Lords Sidmouth (the former PM, Addington) and Castlereagh, to be displayed on pikes on the steps of the Mansion House, EC4. When the authorities heard of the plot, they sent police and soldiers into Cato Street on the 23rd. The loft was stormed and most of the conspirators captured, but Thistlewood killed one of the police officers and fled. He was caught at no. 8 White Street, Little Moorfields. The conspirators were imprisoned in the Tower of London, Thistlewood himself in the Bloody Tower. Five of them including Thistlewood, were hanged outside Newgate Prison, EC4 (the site is now the Old Bailey) on 1 May.

London has been the scene of other revolutionary schemes, the most famous was the Gunpowder Plot, by a group which included Guy Fawkes. This was an attempt to blow up Parliament and the King on 5 November 1605. The plot was discovered, as is recorded in a plaque at 244-278 Crondall Street, Hoxton N1: 'In a house near this site on the 12th October 1605 Lord Monteagle received a letter unmasking the plot led by Guy Fawkes to blow up the Houses of Parliament'. Fawkes and the others were tried in Westminster Hall and executed on 1 February 1606 in Old Palace Yard, Westminster SW1.

More recently, the Conservative Cabinet was the target of an IRA mortar attack on No. 10 Downing Street on 7 February 1991, during the Gulf War. No one was injured. The Prime Minister, John Major, said, 'I think we'd better start again somewhere else' and the meeting adjourned to the secure 'Cobra Room' underground. The mortars were fired from a van parked at the corner of Horse Guards Avenue, next to the Banqueting House.

SMOKE-FILLED ROOMS

Political history is full of party leadership plots. For example, just two days after the sudden death of Labour Party leader, Hugh Gaitskell in January 1963, Tony Crosland's top-floor flat in the Boltons, Chealsea SW10, was the scene of a meeting of those in the party determined to stop Harold Wilson from winning the leadership election. Further meetings took place in the flat of Jack Diamond in Greycoat Place, Westminster, SW1. There was a similar meeting took place in the house of the Tory junior minister, Tristan Garel-Jones at 12 Catherine Place, Westminster,

SW1 following the failure of Margaret Thatcher to win re-election as party leader on the first ballot in November 1990.

At the height of the Tory leadership election in the summer of 1995, journalists spotted BT engineers outside 11 Lord North Street, SW1, apparently installing new telephone and fax lines, and several Tory notables were seen entering the house. The rumour immediately spread that the house was being prepared for a leadership bid by the cabinet minister, Michael Portillo, which he refused to confirm or deny at the time.

Political Women

From the suffragettes onwards London has been a centre of feminist movements and radical women have left their mark on London's landscape.

SUFFRAGETTES

A notable example of women's political activity in modern times is perhaps the suffragist (or suffragette) movement, and there are plenty of relevant sites in central London.

The headquarters of the Women's Social and Political Union, founded by Emmeline Pankhurst, were at no. 4 Clement's Inn, off the Aldwych, WC2, in 1907. There is a memorial to the suffrage movement in the north-west corner of Christchurch Gardens, Victoria Street, SW1, known as the Suffragette Scroll (1970), which has a stirring inscription.

Some sites reflecting suffragist action are 10 Downing Street, SW1 itself, where three activists tried to disrupt a Cabinet meeting in January 1908, two of them by chaining themselves to the railings. Following a similar incident in March 1912, when stones were thrown at the windows of No. 10, some demonstrators were arrested and held at the former police station in nearby Cannon Row (now Canon Row) SW1.

In a cupboard in the Crypt Chapel under Westminster Hall, Emily Wilding Davison, hid for 48 hours in 1911 so that she could put 'House of Commons' as her address in that year's census. In 1989–90 Tony Benn erected a plaque in the small room to mark her protest against the then Liberal Government's failure to legislate for women's suffrage. Davison later became the movement's most famous martyr when she fatally threw herself under the King's horse during the Epsom Derby in 1913.

Leaders of the WSPU were the Pankhursts, Emmeline and her daughters, Christabel and Sylvia. Sylvia lived at 120 Cheyne Walk, Chelsea, SW10 and at nearby 45 Park Walk. She set up an East London toy factory and babies' nursery at 45 Norman Grove, E3, marked by a plaque. Emmeline is buried at Brompton Cemetery, Old Brompton Road, West Brompton SW6. There is an impressive memorial to her, and to Christabel, in the Victoria Tower Gardens, Millbank, SW1, which was unveiled by Stanley Baldwin in 1930.

Caxton Hall, Caxton Street, SW1 was a well-known venue for many suffrage meetings (the Suffragette Scroll is only yards away), including one organised by the Pankhursts in February 1906, on the day of the State Opening of Parliament. When they heard that the King's Speech did not contain any promise of women's suffrage, they marched on Parliament in protest.

Hyde Park was the scene of a huge meeting of 50,000, on 26 July 1913, marking the culmination of a six week Women's Pilgrimage. Also in Hyde Park is the Reformer's Tree, a spot (nowadays marked by a lamp post) where an oak tree once grew, which was a rallying point for the Reform League and other protest groups, and weekly meetings of the WSPU were held near it.

LEADING FEMINISTS

Dame Millicent Fawcett, president of the National Union of Women's Suffrage Societies from 1897 to 1919, and an opponent of the Pankhursts' more militant approach, lived at 2 Gower Street, Bloomsbury, WC1 (where there is an LCC plaque), where she died in 1929. There is a memorial to her in Westminster Abbey.

Three residences of Mary Wollstonecraft, a key feminist writer of the late 18th century, are at 209-215 Blackfriars Road, Southwark, SE1; 373 Mare Street, Hackney, E8, and Oakshott Court, Werrington Street, Somers Town, NW1. She was originally buried in Old St Pancras Churchyard, NW1, where her headstone remains.

PARTY POLITICALS

Countess Constance Markievicz was interned in Holloway Prison for almost a year in 1918 to 1919. During this time she became the first woman to be elected to the House of Commons, as a Sinn Fein MP for Dublin, but she refused to take her seat. So the honour of the first woman to sit in Parliament (as a Conservative in 1919) went to Nancy Astor. Lady Astor was married in All Souls Church, Langham Place, Marylebone, W1 on 3 May 1906, and lived at Astor House, 2 Temple Place, WC2, and at 4 St James's Square, SW1, where there is a plaque.

Margaret MacDonald, the wife of the first Labour Prime Minister, Ramsay MacDonald, was an active social and political reformer. She was born at 17 Pembridge Square, Bayswater, W2 and died at her home at 3 Lincoln's Inn Fields, Covent Garden, WC2 in September 1911 when she was only 41. There is a bronze memorial seat to her on the north side of the Fields, by R R Goulden, which was unveiled in December 1914. It shows her surrounded by nine children, and the inscription records that she "spent her life in helping others."

The socialist writer Vera Brittain (mother of Shirley Williams) is commemorated by plaques at two addresses, 11 Wymering Mansions, Wymering Road, Maida Vale, W9 and 58 Doughty Street, off Gray's Inn Road, WC1.

Britain's first woman Prime Minister, Margaret Thatcher, has lived in a number of London addresses, including Swan Court, Chelsea; Eaton Square, Belgravia; Westminster Gardens, Marsham Street, SW1, and St George's Square Mews, Pimlico. She married Denis Thatcher at Wesley's Chapel, City Road, EC1 on 13 December 1951. The office of the Margaret Thatcher Foundation is at 76 Chesham Place, Belgravia.

Strangers in a Strange Land

It is not surprising that such a major capital city has been home to many notable people from abroad. Some have come as refugees or exiles, perhaps from domestic oppression, or imigrants in search of a way out of poverty. Some of these people's London sojourns are marked by plaques, statues and other memorials.

BENJAMIN FRANKLIN

Franklin left his mark across polit- ical and social London. He lived for many years at 7 (now 36)

Craven Street, off Charing Cross WC2, lodging with son, William, in the house of a Mrs Stephenson. It nearly burned down in 1762, as he recorded: 'Our house and yard were covered with falling coals of fire, but, as it rained hard, nothing catched' (LCC plaque). He worked in the printing shop in the Lady Chapel of St Bartholomew the Great, St Bartholomew Close, Smithfield, EC1 in 1724.

MAHATMA GANDHI

A leader of non-violent Indian nationalism, he spent his first night in London in 1881 at the Victoria Hotel, Northumberland Avenue, WC2, and lived at 20 Baron's Court Road, Hammersmith, W14, while a law student at the Inner Temple. He also lived in Kingsley Hall, 21 Powis Road, E3, a centre for the poor of the East End. Tavistock Square, WC1 has a statue of Gandhi at its centre, unveiled by Harold Wilson in 1968.

MOHAMMED JINNAH

A founder of Pakistan, he was a law student at Lincoln's Inn. He lived at 35 Russell Road, W14 in 1895, marked by an LCC plaque, and bought a house in West Heath Road, Hampstead, NW3.

GIUSEPPE MAZZINI

A leading Italian nationalist, he lived at 183 North Gower Street, Bloomsbury, NW1 (1837-40), and studied at the British Museum. He also lived for a number of years at 18 Fulham Road, SW6. There are two other plaques in central London honouring him as an apostle of Italian democracy. Both were erected privately. One is at 5 Hatton Garden, EC1, and the other at 10 Laystall Street, EC1, where he founded the Italian Operatives' Society in defence of Italian workers. It became the Mazzini and Garibaldi Club, later based at 51 Red Lion Street, EC1.

KARL MARX

His first London house was 4 Anderson Street, Chelsea SW3, from which the family were evicted for non-payment of rent. They moved to the German Hotel, 1 Leicester Street, WC2 (later the site of Manzi's, a fish restaurant), and then for six years in nearby Dean Street, Soho, W1. Six months were spent at no. 64, and from 1850-56 in two rooms on the top floor at no.28, above the later Leoni's Quo Vadis Restaurant.

The choice of the latter residence as the site of a blue plaque was not uncontroversial. Peppino Leoni, the downstairs restaurateur, was outraged: 'My clientele is the very best ... rich people .. nobility and royalty — and Marx was the person who wanted to get rid of them all!'. Nevertheless, a GLC plaque was erected there. It is not known what effect it had on the local economy.

The family then spent eight years at no. 9 (later no. 46) Grafton Terrace, Gospel Oak, NW1, and then in nearby Maitland

Park Road. They lived first at no. 1 (then 1 Modena Villas) and, later, for the last eight years of Marx's life, at no. 41 (rebuilt, after being bombed in the last war, as nos. 101-108).

Among the many sites in central London connected with Marx and his circle are 18 Greek Street, Soho, W1, the first meeting place of the International Working Men's Association, and St Martin's Hall, Long Acre, Covent Garden, WC2, site of the First International in September 1864. The Marx Memorial Library is at Marx House, Clerkenwell Green, EC1.

Marx was buried in Highgate Cemetery, Swain's Lane, N6 in March 1883, and was later moved to the present site in the East Cemetery in 1954. The famous massive bronze bust of Marx was placed there in 1956. Other members of the Marx family are also buried there.

VLADIMIR ILICH LENIN

Soviet Communist revolutionary leader, he lived at 16 Percy Circus, south of Pentonville Road, WC1 in 1905, marked by a private plaque on the site of the property. He also lived in a number of places around London, including 30 Holford Square (off King's Cross Road/Great Percy Street) in 1903, a monument opposite unveiled by the Soviet ambassador 1943, and 6 Oakley Square, Camden Town NW1 in 1911.

He pursued his radical political activities in London. In what became the Marx Memorial Library, Marx House, Clerkenwell Green, EC1 was the first office of Twentieth Century Press which printed 17 issues of Iskra (The Spark). The Jewish Social Club Hall in Fulbourne Street, E1 held the 5th Congress of the Russian Social Democratic Labour Party, attended by luminaries such as Lenin, Stalin and Trotsky.

LOUIS NAPOLEON

The future Napoleon III, nephew of Bonaparte and Emperor of France, lived in London over several periods. Among his residences in fashionable St James's, SW1 were 1c King Street, 1 Carlton Gardens, and the Brunswick Hotel, Jermyn Street, (where he adopted the alias of the Comte d'Arenberg). Following his defeat in the Franco-Prussian war in 1870, he returned to Britain, living outside urban London in Camden Place, Chislehurst Common, Chislehurst. He enjoyed London living, being an honorary member of the Army and Navy Club, and a habitué of Crockford's. When he returned as Emperor, and was processing up St James's Street, he was spotted pointing out to his wife 'with interest and pleasure' his former home at 1c King Street.

Others who have spent some time in London include:

JAWAHARLAL NEHRU

India's first Prime Minister lived at 60 Elgin Crescent, W11 from 1910 until 1912 while he was a student at the Inner Temple.

THOMAS MASARYK

Future Czechoslovak president, lived and worked at 21 Platt's Lane, NW3 during World War I.

SIMON BOLIVAR

Latin American nationalist leader, lived at 4 Duke Street, W1 in 1810. A monument to him in Belgrave Square, SW1, was unveiled by James Callaghan (then Foreign Secretary) in 1974.

EDWARD BENES

Czech leader, lived at 28 Gwendolen Avenue, off Upper Richmond Road, Putney, SW15.

FRANCISCO DE MIRANDA

South American nationalist leader, lived at 58 Grafton Way, W1 (1803-10). There is also a statue in adjoining Fitzroy Street.

DAVID BEN GURION

Israeli premier, lived at 75 Warrington Crescent, Maida Vale, W9.

SLOBODAN YOVANOVITCH

Prime Minister of Yugoslavia, lived at 58-66 Cromwell Road, South Kensington, SW7 from 1945 to 1958

LAJOS (LOUIS) KOSSUTH

Hungarian mid-19th century nationalist, lived at 39 Chepstow Villas, Notting Hill, W11.

MARCUS GARVEY

Black nationalist, lived briefly at 2 Beaumont Crescent, Earl's Court, W14.

RAMMOHAN ROY

First ambassador to Britain of the Mogul Emperor, lived at 29 Bedford Square, WC1.

SUN YAT SEN

Chinese revolutionary leader, lived at 4 Warwick Court, Gray's Inn, WC1, and at 49 Portland Place, W1, in 1896.

BERNARDO O'HIGGINS

Chilean nationalist leader, lived at Clarence House, 2 The Vineyard, Richmond.

Riots & Affrays

As Britain's centre of government, London has attracted more than its fair share of political protests, many of which have had enormous historical significance.

PEASANTS' REVOLT, JUNE 1381

A protest against economic conditions and imposition of poll taxes, it was most significant in Kent where it was led by Wat Tyler. His protesters marched on London, meeting up with other groups at Blackheath Gate, Greenwich and elsewhere. The Tower of London itself was occupied, and King Richard II was forced to negotiate with Tyler at Smithfield. However, Tyler was mortally wounded by the Lord Mayor of London, Sir William Walworth, and was carried to St Bartholomew's Hospital, Smithfield, EC1. It is said that Tyler was later dragged out of the building and beheaded. At Fishmongers Hall, London Bridge (north side) EC3, the Fishmongers' Company has the dagger reputedly used by Walworth to stab Tyler. There is a life-size wooden statue of Walworth (1684) on the main staircase.

GORDON RIOTS, JUNE 1780

A week of anti-Catholic rioting and a march on Westminster from St George's Fields, Southwark caused many deaths and injuries throughout London, following Lord George Gordon's petition against Catholic relief legislation. The homes of many leading political figures were attacked. The Prime Minister, Lord North, was threatened by a mob in Downing Street. Churches were attacked, including the Sardinian Chapel, close by Lincoln's Inn Fields, WC2, and the Bavarian Chapel in Warwick Street, Soho, W1. Prisons around London were invaded, and prisoners released; bridges were captured and even the Bank of England was under siege.

Rioters attacked the Bloomsbury home of Lord Mansfield, the Lord Chief Justice, but, on finding their quarry not there, went to his Kenwood house in Hampstead. They stopped en route at the Spaniard's Inn, Spaniard's Road, NW3, where the landlord managed to delay them with offers of drink until the military arrived and captured them.

When an inventory was taken in 1782 of the contents of the Grosvenor Square home of the Whig leader, the Marquess of Rockingham, it was found to include 'an iron bar taken from one of the rioters in June 1780'.

THE BATTLE OF CABLE STREET, OCTOBER 1936

Locals tried to prevent a march by Oswald Mosley's British Union of Fascists through an area in Whitechapel, E1, originally

Gardiner's Corner at the junction of Whitechapel Street and Leman Street. The marchers re-routed through Cable Street, passing the junction with Christian Street. Anti-fascist protesters fought with police. These events led to the passing of the Public Order Act 1936. There is a mural of the 'Battle' In St George's Town Hall, 236 Cable Street.

Scandals

L ondon has inevitably had its share of political scandals over the years. Some examples are included below:

- Gladstone's curious preoccupation with 'fallen women' in various parts of London (including Haymarket, SW1 and Great Windmill Street, W1).

- Charles Stewart Parnell's catastrophic affair with Kitty O'Shea, whom he met at a dinner party at the O'Sheas' Victoria flat in 1880, and with whom he stayed in Wonersh Lodge, Eltham, then a suburban village in south-east London.

- The affairs (at his flat at 76 Sloane Street, Belgravia, SW1) which destroyed the Cabinet career of Sir Charles Dilke, MP for Chelsea, in the mid-1880s.

- The alleged blackmailing of Ramsay MacDonald by an Austrian woman in Horseferry Road, SW1, over erotic love poetry she claimed he had sent to her.

Two fraudsters

- Horatio Bottomley learned the tricks of his later trade as a serial swindler when he was an office boy at a solicitors' firm in Coleman Street, off Moorgate, EC2. He had a line of failed publishing ventures, leading to bankruptcy, the loss of his Pall Mall flat and the loss, after only two years, of his parliamentary seat in 1912. He became an MP again six years later, but resumed his career of financial criminality, and ended up in jail. He died in 1932, following a heart attack at the Windmill Theatre.

• J. Maundy Gregory, a key fixer in the 'honours for sale' affair during Lloyd George's premiership, lived at 10 Hyde Park Terrace, W2. He operated from a suite of offices at 38 Parliament Street, SW1, where he went to great lengths to avoid detection by varying his daily route to work, and by always using the back door onto Canon Row.

THE PROFUMO SCANDAL

This had everything, including sex and spies, and when it blew open in 1963, it hastened the demise of Harold Macmillan's premiership a few months later.

The scandal revolved around the activities and friends of an osteopath, Stephen Ward, who worked at 38 Devonshire Street W1. He lived at 17 Wimpole Mews, Marylebone, W1, where Christine Keeler (who had a flat at 63 Great Cumberland Place, Marylebone, W1) met the Soviet naval attaché, Yevgeny Ivanov and the Secretary of State for War, John Profumo (who lived at Chester Terrace, on the east side of Regent's Park, NW1) at different times. Ward had to leave his home and move to a flat at Bryanston Mews, W1, which once belonged to the notorious landlord, Peter Rachman. Various entertainment venues featured in the scandal, from the Rio Café, Westbourne Park Road, Notting Hill, W11, to an 'All Nighters Club' in Wardour Street, Soho, W1.

Two 18th Century Giants

John Wilkes was born in 1725 in St John's Square, Islington, EC1. He married an heiress, Mary Meade, in 1747 in St John's Church, St John's Square, EC1. Wilkes was an MP until 1790, and also Lord Mayor of London in 1774. Issue 45 of his journal, the *North Briton*, in April 1763, attacking the King's speech proroguing Parliament, led to prolonged legal battles with the authorities. He also fought a duel with a government supporter, Samuel Martin, in November 1763 in Hyde Park, receiving a serious bullet wound in the stomach.

He had many addresses throughout central London, including 13 Great George Street, SW1 until 1764; 7 Princess Court, Westminster until 1790; St James's Place, SW1, and 30 Grosvenor Square, W1. He also resided at 2 Kensington Gore, SW7, where he set up his lover and their daughter. The Tower of London was also his 'residence' for a time in 1763, as was the King's Bench Prison, near St George's Fields, Southwark. Wilkes was imprisoned in the latter after winning the Middlesex election in 1768. His supporters demonstrated outside the prison, and on 10 May around 15,000 assembled crying 'Wilkes and Liberty!' They were shot at by fearful troops, killing seven demonstrators, in what became known as the 'Massacre of St George's Fields'.

Wilkes died in December 1797 in his Grosvenor Square house and was buried in a vault in the nearby Grosvenor Chapel, South Audley Street, Mayfair, W1.

There is a famous statue of Wilkes at the corner of New Fetter Lane and Fetter Lane, to the north of Fleet Street, EC4. Unveiled in October 1988, the statue displays his squint, thereby being claimed as the only cross-eyed statue in London.

Charles James Fox was born on 24 January 1749 in Conduit Street W1, where the family was temporarily in residence while their usual London home, Holland House (in what is now Holland Park) was being repaired. He had a remarkably varied political career, several times as a leading government minister from the 1770s up to his death in 1806. However, he is perhaps more famous as an opposition champion of parliamentary reform. Like Wilkes, he had a duel in Hyde Park, when he faced William Adam in November 1779.

His election campaigns famously dramatic, especially the 1784 campaign for the Westminster constituency. The portico of St Paul's Church, in Covent Garden, was frequently used for election hustings. The celebrated Georgiana, Duchess of Devonshire was an ardent Foxite, and was reputed to campaign for Fox by offering kisses to voters in the streets of Covent Garden.

Fox had many London addresses. He was based at 46 Clarges Street, Piccadilly, W1 in 1803-4 while campaigning against Addington's Tory Government, and the site (currently the 'Fox Club'), has a plaque. He also lived at 26 (now 9) South Street, Mayfair, W1 (later the site of Egyptian Embassy offices); in St James's Street, next to Brooks's Club, where he incurred huge gambling debts. At Almack's Club, 50 Pall Mall, SW1, he once played faro for eight hours, losing £11,000. When friends found him afterwards, quietly reading Herodotus in Greek, he explained, 'What else is there to do when a man has lost every-thing?'

Fox died in September 1806 in the Duke of Devonshire's villa, Chiswick House, Burlington Lane, W4, and was buried in the North Transept of Westminster Abbey, SW1 (where there is a memorial). There is a statue of Fox as a Roman senator holding the Magna Carta, in Bloomsbury Square, WC1, erected in 1816.

METROPOLITY

City Links A-Z

ALBERT HALL

The site at Kensington Gore, SW7 was once occupied by Gore House where William Wilberforce, the slavery abolitionist, lived in the early 19th century. Later, as the residence of Lady Blessington and Count d'Orsay, between 1836 and 1849, it became a major society centre. Wellington once visited and was delighted by a talking crow which said 'Up boys and at 'em'.

The Albert Hall was opened by Queen Victoria in March 1871 in the presence of Gladstone and Disraeli. It has been the venue of many political events, such as major speeches by the Tory leader, Arthur Balfour, during the second 1910 election, and two years later by his successor, Bonar Law, strongly attacking the Liberal Government. The 1929 Conservative Conference was also held there. In the early 20th century, political meetings were often interrupted by supporters of women's suffrage; one even hid in an organ pipe and sent strange wailing noises through a microphone.

BIRTHPLACES OF PARTIES

- Willis's Rooms (formerly Almack's Assembly Rooms, later Almack House) in King Street, St James's, SW1 was the scene of a famous meeting on 6 June 1859 of various non-Tory parliamentary factions, popularly deemed to be the birth of the Liberal Party.

- In Farringdon Street, EC4 is the site of the Congregational Memorial Hall where the Labour Representation Committee was formed on 27 February 1900 (marked by a GLC plaque), and where, at a conference following the general election in early 1906, the LRC became the Labour Party.

- David Owen's home at 78 Narrow Street, Limehouse, E14 was the location for the so-called 'Limehouse Declaration' by several leading dissident Labourites on 25 January 1981. This led on 26 March that year to the formation of the Social Democratic Party, which was formally launched at the Grand Hall of the Connaught Rooms, Great Queen Street, WC2.

BUCKINGHAM PALACE

The official London residence of the Sovereign since the mid-19th century. The Prime Minister has a weekly audience, and other government ministers can be seen entering and leaving the central gateway, especially when being appointed or removed from office.

Two interesting connections between the Sovereign and the government and Parliament concern the Vice-Chamberlain of the Household. This is a traditional Royal Household office held by one of the Government Whips. During the State Opening of Parliament the Vice-Chamberlain is 'held hostage' at Buckingham Palace, to ensure the Sovereign's safe return from Westminster. The Vice-Chamberlain writes a daily letter to the Sovereign, whenever Parliament is sitting, with all the relevant news of that day's parliamentary business.

CALL OF DUTY

In order that as many MPs and peers as possible can vote in parliamentary divisions, many sites near to the Palace of Westminster have a 'division bell' for the House of Commons and/or the House of Lords installed on their premises. To have such a facility is often regarded as quite a status symbol, and in order to be permitted to have one, an applicant must have the written support of six Members of Parliament.

Other than obvious locations such as government offices, party HQs, conference halls and private residences, many restaurants, pubs and hotels in the Westminster area have a division bell. These include the St Ermin's Hotel, Caxton Street; the St Stephen's Club, 34 Queen Anne's Gate; Shepherds Restaurant, Marsham Court; Albert Carvery Restaurant, 52 Victoria Street; Ritz Hotel, Piccadilly; Footstool Restaurant, St John's Smith Square, and the Royal Horseguards Hotel, Whitehall Place.

CITY SPEECHES

The heart of the City of London hosts annual speeches by senior government ministers:

• The Guildhall is the seat of the City of London's local government, the City Corporation, holding meetings of its Court of Common Council. The Great Hall is the scene of major gatherings, and boasts statues of Chatham, Pitt, Wellington and Churchill. It is also the venue for the annual Lord Mayor's Banquet each November, at which the Prime Minister traditionally makes a major speech.

• The Mansion House is the scene of the Chancellor of the Exchequer's speech at the Lord Mayor's annual dinner each June. One of the most famous was that of Lloyd George in 1911, which he used to warn Germany about its aggressive foreign policy, sparked by its sending of a gunboat to the Moroccan port of Agadir. The Foreign Secretary also makes a major speech at the Lord Mayor's Easter Banquet.

DUELS

In earlier times, political disputes would sometimes be settled by a duel. Some favoured sites were:

- Battersea Fields (now Battersea Park), SW11: The Duke of Wellington fought a duel with the Earl of Winchelsea over Catholic Emancipation on 21 March 1829. No-one was injured and honour was satisfied.

- Hyde Park: The Earl of Shelburne duelled with a Lt–Col Fullarton on 22 March 1780. He was wounded in the groin, but the affair made him very popular.

- Putney Heath, SW15: Pitt the Younger fought a duel on 27 March 1798 with a leading Whig, George Tierney, over a heated exchange in the Commons. Neither was harmed. The famous duel between two Cabinet ministers, George Canning and Viscount Castlereagh, also took place here, at 6am on 21 September 1809, over personal rivalry and jockeying for ministerial office. Canning was shot in the thigh.

- St James's Park: A favourite spot for duels in 18th century. William Pulteney and Lord Hervey duelled with swords on 25 January 1730, over the former's attacks on the latter's defence of Walpole

IRISH TROUBLES

London has suffered from many incidents connected with the history of Ireland. Bombs have caused many casualties and much damage, including recently Chelsea Barracks (October 1981), Hyde Park and Regent's Park (July 1982), Harrods (1983), Bishopsgate (April 1993) and Canary Wharf (February 1996).

36 Eaton Place, SW1 saw the murder of Sir Henry Wilson MP, former Chief of the Imperial Staff in 1920s by Irish terrorists on 22 June 1922. There was also an attempt on Edward Heath, when a two pound bomb was thrown from car outside his Wilton Street, Belgravia home, where he lived for a time after losing the premiership in 1974.

At Clerkenwell Prison, Clerkenwell Close, EC1, there was a failed attempt to free two Fenians in December 1867 by blowing up the prison wall. The explosion killed six locals. Michael Barrett was found guilty, and executed outside the walls of Newgate Prison (now Newgate Street, EC1), the last public execution in England.

LOCAL GOVERNMENT

As well as being a national capital, London is a city with a long history of municipal government. Outside the confines of the City itself (governed by the Corporation of London), there was a ramshackle system of local government for centuries as the metropolis grew. The London County Council was set up in the late 19th century (based at Spring Gardens, off Trafalgar Square, SW1), and in the 1960s was replaced by the Greater London Council and the 32 boroughs. From 1922 to 1986 the LCC and GLC were based at County Hall, SE1, a convenient location for the GLC's campaign against abolition in the mid-1980s, when the frontage was used for huge billboard messages to the parliamentarians at the Palace of Westminster just across the river.

Borough government in London has always been a lively affair. Poplar in East London became a byword for municipal socialism falling foul of the letter of the law in the 1920s, under the leadership of George Lansbury.

LONDON 'WHITE HOUSE'

Several US Presidents have connections in London other than in Grosvenor Square. John Quincy Adams, son of John Adams, was married at All Hallows by the Tower, St Dunstan's Hill, EC3. Martin van Buren lived at 7 Stratford Place, off Oxford Street, W1, in 1831–32. From 1902 to 1918 Herbert Hoover lived at 39 Hyde Park Gate, South Kensington, SW7.

Teddy Roosevelt married his second wife, Edith Kermit Carow, at St George's, Hanover Square W1 on 2 December 1886, describing himself as a 'ranchman'. They honeymooned in Brown's Hotel in Dover Street, W1, (as did his namesake, Franklin, in 1905).

John F. Kennedy lived at 14 Princes Gate, Knightsbridge SW7, when his father, Joseph, was the US ambassador. There is a memorial bust of JFK at the International Students Hostel, 1 Park Crescent, W1.

Other memorials to presidents include the George Washington statue in front of the National Gallery, Trafalgar Square, presented by the State of Virginia in 1921. There are statues of Abraham Lincoln in Parliament Square, presented by the American people in 1920, and at the Royal Exchange, City, EC3. FDR shares one of London's most amusing recent memorials, the sculpture ('The Allies') of Churchill and Roosevelt chatting on a bench in New Bond Street, W1.

MILITARY MANOEUVRES

One of the most famous political meetings in English constitutional history took place at St Mary's Church, Putney High Street, SW15, in October-November 1647. Cromwell's New

Model Army was based in Putney, and the Army Council used to discuss the future of the country around the communion table with their hats on.

POLITICAL DRINKING

Taverns and coffee houses (most long gone) were popular venues for meetings of political societies and clubs. The British Coffee House, 27 Cockspur Street, SW1 was the location for meetings of various political clubs, such as the Portland Club in 1791, and the Fox Club in January 1813. The Sutherland Arms tavern, May's Buildings, May's Court, WC2, was the meeting place of the Eccentrics Club, whose members included Sheridan, Fox and Melbourne. The Crown & Anchor pub in the Strand, WC2, saw the first London meeting of the famed Liberal reformer, John Bright, in 1842. The King's Head, Bowling Green Lane, Islington, EC1, was the venue for the first meeting of the London Patriotic Society in 1869. Meetings of the October Club, a Tory club, were held in the Bell Coffee House in King Street, Westminster, SW1.

At least two mock parliamentary clubs flourished in coffee houses. At the Crown Coffee House in Drury Lane, WC2, in the mid-18th century, were held meetings of the 'Flash Cove's Parliament' where each member took the title of a member of the Lords. The 'House of Lords Club', in the first half of the 19th century, started in the Fleece Tavern, Cornhill, EC3, then at the Three Tuns, Southwark and the Abercrombie Coffee House, Lombard Street, EC3. Its members were not peers, though they took pretend titles for themselves, and meetings were presided over by the 'Lord Chancellor' wearing legal wig and robes. There was a sliding scale of fees by title, e.g. 1/- for a baron, 5/- for a duke.

QUEEN'S HALL, LANGHAM PLACE

A major venue for political meetings of all kinds, including several party conferences, such as Labour in June 1923 and October 1924, and the Conservatives in 1912 (twice), 1922 and 1930.

It was here on 18 March 1931 that Stanley Baldwin, the Tory leader, attacked the Press lords over their campaigns against his policies, using a phrase suggested by his cousin, Rudyard Kipling. He described them as wanting "power without responsibility, the prerogative of the harlot throughout the ages".

ROYAL ROUTE

When the Sovereign goes to the Palace of Westminster for the State Opening of Parliament, the procession takes the following traditional route: Buckingham Palace – The Mall – Horse Guards Parade – Horse Guards Arch – Whitehall – Parliament Square – Palace of Westminster (Sovereign's Entrance). The State Opening usually takes place in November each year, although this may vary after a general election.

SOMEWHERE PRIVATE

Sometimes a private house is a convenient and sufficiently secret venue for especially sensitive meetings. For example, 96 Cheyne Walk, Chelsea, SW1 home of the then junior Northern Ireland minister, Paul Channon, was used for a meeting on 7 July 1972 between IRA and British Government representatives.

TOWER OF LONDON

Various parts of the Tower have been used as a prison for 'enemies of the state'. After the 1660 Restoration, some of the accused regicides of Charles I were imprisoned here. Some died, while others were taken elsewhere and executed. Colonel Thomas Harrison, who had played a significant part in the trial and sentencing of the King, was executed at Charing Cross. Similar fates befell leading Jacobites including, in April 1747, Lord Lovat, the last person to be beheaded in England. More recently Sir Roger Casement was held in St Thomas's Tower in 1916 before his treason trial, and Rudolf Hess was detained in the Governor's House for several days following his flight to Britain in 1941.

Two prime ministers had very different experiences of the Tower. In January 1712, Robert Walpole, accused of "high breach of trust and notorious corruption" when Secretary at War, was expelled from the Commons and imprisoned briefly in the Tower. While detained he still managed to receive many visitors. On the other hand, the Duke of Wellington was regarded as an active, successful Constable of the Tower (1824-52). He had the moat drained and the Wellington barracks built.

TRAFALGAR SQUARE

In the 19th century, this Square was the place for the airing of the grievances of parliamentary and social reformers. In the 20th century, the causes ranged from unemployment and women's suffrage to nuclear disarmament.

A demonstration by the unemployed on 8 February 1886 led to disturbances known as 'Black Monday'. On 13 November 1887 there was rioting following a protest by radical socialists and others, which became known as 'Bloody Sunday'. More recently, the Square saw some of its most intensive activity during the demonstrations against the introduction of poll tax in March 1990.

VARIOUS VENUES

- Caxton Hall, Caxton Street, SW1, was the site of many suffragette rallies, and of Liberal Party Assemblies in 1942 and 1950. At a Conservative Party meeting on 30 November 1930 to discuss policy on Empire Free Trade, the leader, Stanley Baldwin, said to the paparazzi as he went in: "photograph me now gentlemen; it may be the last time you will see me". In the register office, Anthony Eden was married to a niece of Winston Churchill in August 1952. Churchill himself is commemorated by a plaque for speaking at the Hall between 1937 to 1942.

- Clerkenwell Green, EC1, was a favourite place for large meetings and demonstrations. William Cobbett addressed an anti-Corn Law meeting there in February 1826 and Fergus

O'Connor spoke at a Chartist meeting in April 1848, a week before the famed Chartist march from Kennington Common.

- Conway Hall, Red Lion Square, WC1, is the home of the South Place Ethical Society and a popular venue for political events and meetings, often those not directly in the political mainstream.

- Kingsway Hall, 70 Great Queen Street, WC2, was the scene of a speech by the future Pakistan leader, Mohamed Jinnah, to the Muslim League in 1917, and a special Conservative Party Conference was held there in 1946.

- Speaker's Corner, Hyde Park, near Marble Arch, W1, is a traditional spot for anyone to speak on any subject. The right of assembly was established in 1872 as a result of public demonstrations mounted by the Reform League to campaign for a wider vote.

- St Ermin's Hotel, Caxton Street, SW1 has long been a popular meeting place for politicians because of its proximity to Westminster and Whitehall. The hotel has a division bell, and there is a passageway, now blocked up, that ran from under the main staircase to Parliament. It saw the launch of Conservative 'Yes' Campaign during the Common Market Referendum, and a two day meeting of Liberal MPs in June 1977 to debate the continuation of their agreement with the then minority Labour Government, the 'Lib–Lab pact'.

- St James's Hall, Piccadilly, SW1, on the site of the later Piccadilly Hotel was the venue of the 1897 Conservative Party Conference.

WELLINGTON CORNER

There is so much information about the Duke of Wellington around Hyde Park Corner that it deserves to be renamed. The Wellington Arch dominates the roundabout, and his former home, Apsley House ('No. 1, London') is now the Wellington Museum. The pedestrian underpasses are lined with tiles telling his story, and the Achilles statue just inside Hyde Park commemorates his victories.

Political Eating, Drinking and Shopping*

Westminster is somewhat of a culinary desert – at least that's its reputation. But there are a few oases in the desert, where one can sample anything from ordinary British pub food to the most exquisite five-star menu on offer in London.

RESTAURANTS & BARS

- The Atrium, 4 Millbank, London, SW1 (tel 020 7233 0032) is a top class restaurant located within the building that houses all the media organisations covering Parliament. As its name suggests it is situated in the central atrium of the whole building. Delightful airy surroundings together with the political buzz that makes Westminster such a lively place to work in. You're bound to spot a famous politician lunching with a political journalist, or a lobbyist schmoozing with a Minister. The food is excellent, if rather pricey (reckon to pay around £30-40 per person). But be prepared for some slow, but polite, service.

- Bar Excellence, 1 Abbey Orchard Street, SW1 (tel 020 7222 4707). Basement bar and restaurant just off Victoria Street by

* Compiled by John Simmons

the Department of Trade & Industry. Caters for a hip younger crowd and is decorated in very tasteful bright colours. Primarily a bar which serves excellent (and plentiful portions) of simple food, its restaurant is located on a mezzanine balcony above the bar. A superb atmosphere and great European food. Expect to pay about £20-£30 per head in the restaurant. Well worth a visit.

- Gran Paradiso, 52 Wilton Road, SW1 (tel 020 7828 5818) Popular with Conservative MPs despite the fact that it's quite a hike from the House of Commons - located near to Victoria Station. Wide ranging Italian menu and very reasonably priced.

- L'Amico, 44 Horseferry Road, SW1 (tel 020 7222 4680). Italian basement restaurant, a favourite with MPs and journalists. Walls are decorated with pictures of famous people (mainly politicians and foreign dignitaries) who have lunched there, including Mikhail Gorbachev!

- Shepherds Restaurant, Marsham Court, Marsham Street SW1 (tel 020 7834 9552). Probably the most exclusive and expensive restaurant in the Division Bell area surrounding Parliament. High class English fayre, so if you like your Roast Beef and Yorkshire Pudding followed by Spotted Dick, this is the place for you. Decor is classic English wood panelling with booths for privacy. Another good place to spot famous politicians.

- Olivo, 21 Eccleston Street, SW1 (tel 020 7730 2505). One of the best restaurants in London. Exquisite Italian and Mediterranean food. Quite small so advance booking is advisable. Service and presentation is excellent. Expect to pay £30-£40 per person.

- Kundan Indian Restaurant, 3 Horseferry Road, SW1 (tel 020 7834 3434). One of former Prime Minister John Major's favourite eateries. Basic Indian menu, this restaurant is particularly popular with backbenchers on a budget.

- Simply Nico, 48a Rochester Row, SW1 (020 7630 8061). One of London's finest restaurants with a superb (and expensive!) menu. If you want to impress someone (particularly with the size of your wallet) this is the place to take them.

- Politico's Coffee House, 8 Artillery Row, SW1 (tel 020 7828 0010). Small balcony coffee house located above Politico's Political Bookstore. Serves high class coffee and teas as well as sandwiches, quiches and delicious cakes. Ideal place for a snack break and to watch Parliament live on TV.

- Churchills, Whitehall, SW1. Cheap and cheerful sandwich bar located virtually opposite the entrance to Downing Street in Whitehall. Tony Blair sends out for his sandwiches here!

- Red Lion Pub, 48 Parliament Street, SW1. Rough and ready pub which is popular among political journalists and civil servants. Located opposite the entrance to Downing Street.

- The Speaker Pub, Great Peter Street, SW1. Newly refurbished pub, decorated with pictures of various Speakers of the House of Commons.

- Marquis of Granby Public House 41 Romney Street, SW1. Used to be a haven of left wing intrigue, located as it was next to Transport House, the former home of the Labour Party and TGWU. They have both since moved but it still has that conspiratorial feel to it. Nowadays frequented by staff from nearby Conservative Central Office.

- Westminster Arms, Storey's Gate, SW1. Located to the side of the hideous Queen Elizabeth II Conference Centre, this pub is popular with tourists and politicos alike.

SHOPS

- Politico's Bookstore & Coffee House, 8 Artillery Row, SW1 (020 7828 0010). Located just off Victoria Street by the Army & Navy Department Store, this is Britain's only specialist political bookstore. Politico's stocks a huge range of political books, magazines, think tank reports as well as political gift items, memorabilia, cartoon, videos and tapes. There is also a mezzanine coffee house where you can sip fine coffees and watch Parliament live on TV.

- Parliamentary Bookshop, 12 Bridge Street, SW1 (020 7219 3890). Located on the corner of Parliament Square and Whitehall, the Parliamentary Bookshop specialises in Government publications and books on Parliament.

- Church House Bookshop, 31 Great Smith Street, SW1 (020 7898 1301). Specialist religious bookshop. Well worth a visit.

POLITICAL
LONDON
QUIZ

Questions

Q1. Which famous twentieth century nationalist leader worked
(a) as a dishwasher and apprentice pastry-cook in a Haymarket hotel?
(b) as Post Office savings bank clerk?

Q2. Which small central London street was described by Dickens in his novel *Nicholas Nickleby* as 'a street of gloomy lodging houses ... a sanctuary of smaller Members of Parliament ... There are legislators in the parlours, in the first floor, in the second, in the third, in the garrets; the small apartments reek with the breath of deputations and delegates'?

Q3. Where in central London was said to get its name from a king walking his dogs and was the site of a fatal accident when a prime minister fell off his horse?

Q4. (a) Which famous politician was Guilford Street, WC1 named after?
(b) Which central London street was, at one point, proposed to be named after the Duke of Wellington?
(c) Which hotel boasts the Chartwell Suite and Library and Clementine's Restaurant?

Q5. Which future Prime Minister attended the appropriately named Mr Gladstone's Day School near Sloane Square, SW1?

Q6. Which famous twentieth century political cartoonists lived at the following addresses, both marked by an English Heritage plaque:
(a) 33 Melbury Court, Kensington High Street, W8?
(b) Welbeck Mansions, New Cavendish St, W1?

Q7. What do the following London suburbs have in common: Orpington, Leyton, Bermondsey and Mitcham & Morden?

Q8. In what Belgravia square lived politicos as diverse as Metternich, Tory premiers Baldwin, Chamberlain, Heath and Thatcher, and Liberal PMs Russell and Campbell Bannerman, and was for a time the temporary home of the Speaker following the destruction by fire of the old Palace of Westminster in 1834?

Q9. Who or what was 'Selsdon Man'?

Q10. Where can you see large concentrations of politicians in one place:
(a) in oils?
(b) in wax?
(c) in stone?
(d) on mugs, badges, cards etc.?

Answers

Q1. (a) Ho Chi Minh, the future Vietnamese revolutionary leader, worked in the Carlton Hotel, Haymarket, SW1 (later New Zealand House) in 1913, in the kitchen of the celebrated French chef, Escoffier. A plaque erected by the British Vietnam Association marks this surprisingly domestic piece of history.

(b) Michael Collins, the Irish nationalist, was a clerk for the West Kensington PO Savings Bank while living at 5 Netherwood Road, south of Shepherd's Bush, W14 in the early years of last century, where there is a commemorative plaque by Kensington & Chelsea Council.

Q2. Canon Row, Westminster, SW1. It runs north from Bridge Street, opposite the Palace of Westminster, and is now no longer a public thoroughfare, being flanked by parliamentary offices, including the new Portcullis House. It used to be called Cannon Row, but in recent years reverted to its original spelling.

Q3. Constitution Hill, SW1. The name probably comes from Charles II's constitutional walks with his spaniels. Sir Robert Peel was thrown from his horse here in 1850 and later died from his injuries.

Q4. (a) Lord North, the Conservative PM of the late 18th century, who became the 2nd Earl of Guilford in 1790.

(b) Regent's Park. There were suggestions in the press that it be named after Wellington, and a villa was designed and exhibited at the Royal Academy the following year where he might live in the Park.

(c) The Churchill Inter-Continental Hotel, 30 Portman Square, W1. It also has a Churchill Bar and Divan.

Q5. Harold Macmillan, from 1900-1903. He was born, and lived as a child, at nearby 52 Cadogan Place.

Q6. (a) David Low
(b) Victor Weisz, better known as 'Vicky'.

Q7. They were all the sites of significant by-elections in the second half of the last century. In Orpington in March 1962, the Liberal Eric Lubbock captured what was a Conservative stronghold with a huge swing. It was the most sensational by-election result for 30 years, and marked the start of a Liberal revival. Patrick Gordon Walker, who had been made Foreign Secretary in the new Labour Government in October 1964 despite losing his Smethwick seat in the general election, surprisingly failed to secure his return to Parliament in a by-election in Leyton the following January in what was supposed to be a safe Labour seat. He was forced to resign his Cabinet seat, which cut Labour's already slender Commons majority even further. The Mitcham & Morden by-election in June 1982 was the only one to result directly from the creation of the Social Democratic Party. The former Labour MP, Bruce Douglas-Mann, chose to seek re-election following his defection to the SDP, but lost the by-election to the Conservatives, who were enjoying their post-Falklands War revival. The disastrous Labour showing in the June 1983 general election was presaged by its humiliation in the Bermondsey by-election the previous February, where a supposedly solid Labour docklands constituency fell to the Liberal Simon Hughes following a bitter and divisive campaign.

Q8. Eaton Square, SW1

Q9. In January 1970 leading Conservatives met at the Selsdon Park Hotel, Croydon Road, Croydon for a pre-election policy session, producing a programme which was seen by some as a break from the so-called post-war consensus. It led to Prime Minister Harold Wilson's famous reference to 'Selsdon Man' in a speech at a London Labour Party rally at Camden Town Hall, Euston Road, WC1 a few weeks later, an attempt to portray the Conservatives as uncaring right-wing extremists.

Q10. (a) The National Portrait Gallery, St Martin's Place, off Trafalgar Square, WC2 has an excellent collection of paint-ings (and busts) of politicians past and present
(b) Madame Tussaud's, Marylebone, NW1 exhibits life-sized wax statues of many famous political figures
(c) The Parliament Square area, including the Palace of Westminster and Westminster Abbey
(d) Politico's Bookstore, Artillery Row, SW1 (of course!)

INDEX OF
NAMES & PLACES